SOIL SCIENCE AND AGRICULTURAL CHEMISTRY

a division of

NIPA GENX ELECTRONIC RESOURCES & SOLUTIONS P. LTD.

New Delhi-110 034

SOIL SCIENCE AND AGRICULTURAL CHEMISTRY

A COMPETITIVE EXAMINATION BOOK

Dharmendra Singh Yashona, PhD
Research Fellow
ICAR-Indian Institute of Soil Science
Bhopal, Madhya Pradesh, India

Satish Bhagwatrao Aher, PhD
Scientist-B (Environmental Science)
ICMR-National Institute for Research in Environmental Health
Bhopal, Madhya Pradesh, India

a division of
NIPA GENX ELECTRONIC RESOURCES & SOLUTIONS P. LTD.
New Delhi-110 034

a division of

NIPA GENX ELECTRONIC RESOURCES & SOLUTIONS P. LTD.

101,103, Vikas Surya Plaza, CU Block
L.S.C.Market, Pitam Pura, New Delhi-110 034
Ph : +91 11 27341616, 27341717, 27341718
E-mail:newindiapublishingagency@gmail.com
www: www.nipabooks.com

For customer assistance, please contact
Phone: + 91-11-27 34 17 17 Fax: + 91-11- 27 34 16 16
E-Mail: feedbacks@nipabooks.com

ISBN: 978-93-90591-07-7

Composed and Designed by NIPA.

Preface

The idea behind starting the M.Sc. (Soil Science & Agricultural Chemistry) course is to make experts for sustaining soil health for food security and nutritional safety without impairing environmental quality and creating mass awareness about soil and society relationship. It is a 2-year post-graduate degree program in Indian Agricultural Universities and their affiliated colleges that delves into the details of soil study, mainly for suitable uses in fields such as agriculture. The students who wish to get admitted for this course has to undergo a competitive examination. This applies to those who seek admission in higher studies like a PhD. Similarly, those who have completed the course also need to face similar kinds of examinations to get into the service. The numerous available publications on the multiple-choice questions in Soil Science are quite short and do not provide comprehensive coverage of the subject. These publications lack a subject-specific orientation. They are more diverted to agricultural science rather than in-depth soil science which compel students and job seekers to refer to multiple books and other sources while preparing for the competitive examinations. This in turn results in a loss of their valuable time and energy. The lack of complete information in the existing publications on the subject of Soil Science motivated us for writing a comprehensive book with high-quality multiple-choice questions on various aspects of soil science. The present book starts with the soil glossary followed by six main sections namely General Soil Information, Soil Fertility, Soil Microbiology, Soil Chemistry, Soil Physics and Pedology. The distinction in the sections would help students to be more specific toward the potential opportunity which would save their valuable time while preparing for their career. We are very happy to provide the first edition of Competitive Soil Science and Agricultural Chemistry, a multiple-choice question-based book to the students aspiring to make their career in this field.

We wish to express our deep sense of gratitude to Dr. Brij Lal Lakaria, Principal Scientist, ICAR-Indian Institute of Soil Science, Bhopal for his constant encouragement to us. We are also thankful to M/s New India Publishing Agency, New Delhi for their support in compiling, designing and publishing the book. We do hope, this book will be of immense use to the students who wish to make their career in Soil Science and Agricultural Chemistry field.

Though efforts have been undertaken to keep the book free from inaccuracy, mistakes cannot be overruled, therefore the presence of any typographical error may kindly be notified to the authors/publisher. We welcome suggestions and comments of the esteemed readers for the needful improvement in this publication for the second edition. We wish grand success for the readers seeking opportunities in the field of Soil Science.

Authors

Contents

Soil Glossary

Absorption: the processes by which a substance in take into or in clued with on other substance like intake of water by soil or intake of nutrient by plant.

Acid rocks: An igneous rock that contains more than 60 percentage silica and free quartz.

Acid saturation: The proportion or percentage of a cation exchange site occupied by acid cations.

Acid soils: Low pH with high amounts of exchangeable H^+ and Al^{3+}. Occur in regions with high rainfall.

Acid sulphate soils: Soils that are potentially extremely acid (pH < 3.5) because of the presence of large amounts of reduced forms of sulphur that are oxidized to sulphuric acid if the soils are exposed to oxygen when they are drained or excavated.

Active acidity: The activity of hydrogen ions in the aqueous phase of a soil. It is measured and expressed as a pH value.

Adhesion: It is the force of attraction between molecules of different substance. That is the force of attraction between solid surface to liquid surface. A thin film of water is held in soil particles due to this adhesive force.

Adsorption: the process of sorption of chemical element from solution at the surface of soil particles.

Aeolian deposits: Sand dunes and depositions from the materials transported by wind.

Aggregate: Many soil particles held in a single mass or cluster, such as a clod, crumb, block, or prism.

Agriculture: The science of cultivation the soil, producing crops and raising livestock.

Air capacity: The volume of pore space filled with air when the soil is under a tension of 50 milli bar. This value is also called as 'non-capillary porosity'. It corresponds to a pore size of 0.06 mm or larger in diameter.

Air porosity: The proportion of the bulk volume of soil that is filled with air at any given time or under a given condition, such as a specified moisture potential; usually the large pores.

Albedo: An index of the reflection comprising the ratio of reflected radiation to the total incident radiation.

Alluvial fans: Sediments deposited by streams of flowing water when they enter plains from the hilly areas are known as alluvial fans.

Alluvial terraces: The sedimentary deposits that were laid down by the rivers but are not subject to flooding now are called 'alluvial terraces'. Most of the Indo-gangetic plains are alluvial terraces.

Alluvial: The parent materials transported by water sediments that are deposited by the flowing water such as streams, rivers etc.

Aluminosilicates: Compounds containing aluminium, silicon, and oxygen as main constituents. An example is microcline, $KAlSi_3O_8$.

Amendment: Any substance other than fertilizers, such as lime, sulfur, gypsum, and sawdust, used to alter the chemical or physical properties of a soil, generally to make it more productive.

Aminization: It is a process of enzymatic digestion by non-specific heterotrophic bacteria, actinomycetes and fungi by which protein and proteinous compounds are decomposed into amino acids or amines.

Ammensalism: It is a kind of relationship in which a partner suppresses the growth of other partner by producing toxins like antibiotics, harmful gaseous compounds like NH_3, CO_2, ethylene etc.

Ammonification: The biochemical process whereby ammonical nitrogen is released from nitrogen containing organic compounds.

Anion: Negatively charged ion; during electrolysis it is attracted to the positively charged anode.

Available nutrient: That portion of any element or compound in the soil that can be readily absorbed and assimilated by growing plants. ("Available" should not be confused with "exchangeable.")

Available water: The portion of water in a soil that can be readily absorbed by plant roots. The amount of water released between the field capacity and the permanent wilting point.

Bar: A unit of pressure equal to 1 million dynes per square centimeter (10^6 dynes/cm^2). It approximates the pressure of a standard atmosphere.

Base saturation percentage: The extent to which the adsorption complex of a soil is saturated with exchangeable cations other than hydrogen and aluminium. It is expressed as a percentage of the total cation exchange capacity.

Base: A substance that reacts with H^+ ions or releases hydroxyl ions; a substance that neutralizes acid and raises pH.

Beneficiation: In mining and fertilizer production, treatment to improve the quality and handling of pulverized ore by processes such as sieving and flotation to remove unwanted material.

Bioremediation: The decontamination or restoration of polluted or degraded soils by means of enhancing the chemical degradation or other activities of soil organisms.

Buffering capacity: The ability of a soil to resist changes in pH. Commonly determined by presence of clay, humus, and other colloidal materials bulk blending Mixing dry individual granulated fertilizer materials to form a mixed fertilizer that is applied promptly to the soil.

Burning: Severe localized yellowing; scorched appearance.

Bulk density: The mass of dry soil per unit of bulk volume, including the air space. The bulk volume is determined before drying to constant weight at 105°C.

Calcareous soil: Soil containing sufficient calcium carbonate to cause effervescence on the application of a few drops of hydrochloric acid.

Capillary fringe: A zone in the soil just above the plane of zero water pressure (water table) that remains saturated or almost saturated with water.

Capillary water: The water held in the capillary or small pores of a soil, usually with a tension >60 cm of water.

Carbon management: the various practices by which the various from of carbon in our earth can be transformed and stored in suitable from.

Carbon sequestration: the capture and secure storage of carbon from the atmosphere on a long term basis.

Cat clays: Wet clay soils high in reduced forms of sulfur that, upon being drained, become extremely acid because of the oxidation of the sulfur compounds and the formation of sulphuric acid. Usually found in tidal marshes.

Catena: A sequence of soils of about the same age, derived from similar parent material, and occurring under similar climatic conditions, but having different characteristics because of variation in relief and in drainage.

Cation exchange capacity: A measure of the capacity of a soil to hold the major cations: calcium, magnesium, sodium and potassium (including hydrogen, aluminium and manganese in acid soils) and expressed in centimoles of charge per kilogram (cmol/kg) of soil). It is a measure of the potential nutrient reserve in the soil and is therefore an indicator of inherent soil fertility.

Cation: A positively charged ion; during electrolysis it is attracted to the negatively charged cathode.

Chalcophile: Trace elements that show affinity for sulphide minerals are called chalcophiles. This group includes Zn, (Fe), (Mo), Cu and Ni.

Chelate: A type of chemical compound in which a metallic ion is firmly combined with an organic molecule by means of multiple chemical bonds.

Chlorite: A nonexpanding 2:1 aluminosilicate mineral with a sheet of magnesium and hydroxyl atoms in the interlayer space.

Chlorosis: General yellowing of the plant tissue; lack of chlorophyll.

Clay: A mineral particle size less than 0.002 mm diameter. Also, a specific group of aluminosilicate minerals with a high surface area and exchange capacity.

Clod: A compact, coherent mass of soil produced artificially, usually by such human activities as ploughing and digging, especially when these operations are per-formed on soils that are either too wet or too dry for normal tillage operations.

Clogging: A process that blocks pores and reduces water flow in soils. Mechanisms can be physical, chemical, and biological.

Coefficient of linear extensibility: A measure of the potential one-directional change in the volume of a soil as the soil moisture level changes.

Cohesion: Cohesion is the force of attraction between molecules of same substances *i.e.*, between liquid molecules or water molecules. Hence, a thick film of water is formed due to this cohesive force.

Colloid: A material made of organic or inorganic particles with diameter from 0.001 to 0.1 μm.

Colluvium: Unconsolidated, unsorted earth material that has moved downhill under the force of gravity and has accumulated at the base or lower slopes of the hill.

Commensalism: A long-term biological interaction (symbiosis) in which members of one species gain benefits while those of the other species neither benefit nor are harmed.

Compaction: A process of rearranging soil particles to decrease pore space and increase bulk density. It produces a compacted, dense soil layer.

Competition: It is a universal phenomenon of coexistence in a community where one species is suppressed as the two species struggle for the limited supply of a nutrient or other common requirements. Such interaction in soil is harmful.

Conductivity: A measure of the ease with which a material or object conducts or transmits something such as heat (thermal conductivity), water (hydraulic conductivity), or diffusing substances (diffusion coefficient).

Conservation tillage: Any tillage sequence that leaves at least 30 percent crop cover on the soil surface, with the goal of minimizing soil and water loss. It ranges from omitting one or more tillage operations to eliminating all tillage.

Consistency: The interaction of adhesive and cohesive forces within a soil at various moisture contents as expressed by the relative ease with which the soil can be deformed or ruptured.

Creep: In soil science, the slow movement of soil down a slope under the force of gravity. Also, the rolling or pushing of particles along the ground surface by wind.

Cyanobacteria: Chlorophyll-containing bacteria that accommodate both photosynthesis and nitrogen fixation. Formerly called blue-green algae.

Deflocculation: Dispersion, or separation, of clusters of particles into separate individual particles.

Delta deposits: Most of the coarser sediments suspended in the stream water are deposited as alluvial fans, alluvial terraces and floodplains. By the time these streams reach the oceans and seas, the flow velocity is considerably reduced and the finer particles are deposited near the mouth of the river as ‘delta deposits’.

Denitrification: The biochemical reduction of nitrate or nitrite to gaseous nitrogen, either as molecular nitrogen (N_2) or as an oxide of nitrogen (N_2O).

Desorption: the removal of sorbed material from surface.

Diagnostic horizons: Soil Horizons having specific soil characteristics that are indicative of certain classes of soils. Horizons that occur at the soil surface are called epipedons those below the surface, diagnostic subsurface horizons.

Diffusion: Molecular movement along a gradient. Water diffusion occurs from wet areas to dry ones. Gas and solute diffusion occur from zones of high concentration to zones of low concentration.

Dilution effect: When the rate of dry matter production exceeds the rate of uptake of a nutrient, the concentration of that nutrient in the dry matter decreases. This is called 'dilution effect'.

Dioctahedral sheet: An octahedral sheet of silicate clays in which the sites for the six-coordinated metallic atoms are mostly filled with trivalent atoms, such as Al^{3+}

Dosing: In waste disposal, the process of adding waste at intervals to keep the soil aerobic and to reduce clogging.

Edaphology: The science that deals with the influence of soils on living things, particularly plants, including human use of land for plant growth.

Electrical conductivity: The capacity of a sub-stance to conduct or transmit electrical current. In soils or water, measured in dS/m, and related to dissolved solutes.

Eluviation: The removal of soil material in suspension (or in solution) from a layer or layers of a soil.

Endomycorrhiza: A symbiotic association of the mycelium of fungi and roots of a variety of plants in which the fungal hyphae penetrate directly into root hairs, other epidermal cells, and occasionally into cortical cells.

Enzyme: A protein that increases the rate of chemical reactions in an organism's cells.

Epipedon: A diagnostic surface horizon that includes the upper part of the soil that is darkened by organic matter, or the upper eluvial horizons, or both.

Equivalent weight: The equivalent weight of a substance is the weight in grams that in its reaction corresponds to a gram atom of hydrogen or of hydroxyl, or half a gram atom of oxygen, or a gram atom of univalent ion. When one equivalent weight of a substance is dissolved in one litre, it gives 1N solution.

Equivalent: The unit formerly used to describe cation exchange capacity or quantities of ions. One gram atomic weight of hydrogen or the amount of any other ion that will combine with or displace this amount of hydrogen. The amount that provides 1 mole of charge Equals 1 mole.

Essential nutrients: Elements needed by the plant without which it will not be able to survive are called essential nutrients. For an element to be regarded as an essential nutrient criteria, as propounded by Arnon and Stout (1939).

Eutrophication: Nutrient enrichment of lakes, ponds, and other such waters that stimulates the growth of aquatic organisms, which leads to a deficiency of oxygen in the water body.

Exchange acidity: Exchange acidity may be defined as the acidity developed due to adsorbed hydrogen (H^+) and aluminum (Al^{3+}) ions on the soil colloids. The magnitude of this exchange acidity is very high.

Fertigation: The application of fertilizers in irrigation waters, commonly through sprinkler systems.

Fertilizer requirement: The quantity of certain plant nutrient elements needed, in addition to the amount supplied by the soil, to increase plant growth to a designated optimum fabric materials.

Fertilizer: Any organic or inorganic material of natural or synthetic origin added to a soil to supply certain elements essential to the growth of plants.

Field capacity: The percentage of water remaining in a soil 2 or 3 days after it's having been saturated and after free drainage has practically ceased.

Floodplains: During rainy seasons, streams and rivers overflow their banks and flood the surrounding area. The parts of a valley and alluvial plains, which are inundated during, floods, are referred to as floodplains.

Glacial deposits: In polar regions snow continues to accumulate and the pressure of its weight changes the snow to ice and starts moving, in the form of glaciers. The glaciers transport materials embedded in them and also remove the rock protuberances which come in their way. Materials deposited by the glaciers are called glacial deposits. The materials deposited directly by the glacial ice are called glacial till.

Glaciation: The process of geological erosion by means of glacial ice.

Gleying: A process that produces reduction of iron and other elements under conditions of prolonged saturation.

Gravitational water: It is the third stage of soil water where water that moves freely as response to gravity percolates downwards and drains out to deeper layer of soil profile. It is also known as free water. At this condition, the macro and micro pores are completely filled up with water. There is no space for air movement in soil pore spaces. This state will appear when the soil is under saturation.

Greenhouse effect: The theory of "greenhouse effects" was conceived by J.B. Fourler over a century ago. The entrapment of heat by upper atmosphere gases, such as carbon dioxide, water vapour, and methane, just as glass traps heat for a greenhouse.

Gully: A channel cut by the concentrated flow of water during rainstorms. A gully is deeper than a rill and cannot be removed by normal tillage.

Gypsum requirement: The quantity of gypsum required to reduce the exchangeable sodium percent-age in a soil to an acceptable level.

Halophyte: Any plant that grows in saline environments and readily takes up salts.

Hartig net: Ectomycorrhiza association with plant roots with consisting of an intracellular interface of highly branched hyphae forming a latticework between epidermal and cortical root cells.

Heavy metal: A metal having specific gravity of more than 5.0 or having atomic number higher than 20 is termed as a heavy metal. As a corollary, any metal heavier than calcium is a heavy metal. Elements in soils include Cd, Co, Cr, Cu, Fe, Hg, Mn, Mo, Pb, and Zn.

Heavy soil: A soil with a high content of the fine separates, particularly clay, or one with a high drawbar pull, hence difficult to cultivate.

Hematite: Fe_2O_3 a red iron oxide mineral that 'contributes red color to many soils.

Horizons: A layer within the soil profile having morphological characteristics and properties (e.g. colour, texture, and structure) differing from the layer above and/or below it.

Humification: The processes involved in the decomposition of organic matter and leading to the formation of humus.

Humus: The stable, dark colored organic material that accumulates as a by product of decomposition of plant or animal residues added to soil. The term is often used synonymously with soil organic matter.

Hydration: The chemical combination of water with another substance.

Hydraulic conductivity: An expression of the readiness with which a liquid, such as water, flows through a solid, such as soil, in response to a given potential gradient.

Hydrolysis: A reaction that involves products (H^+ or OH^-) of the dissociation of water. A mineral weathering reaction that adds H^+ to a mineral structure.

Hygroscopic water: This is the first stage of soil water content where water is held tightly by the surface of the soil particles by the forces of adhesion or adsorption force. Hence, it is also known as water of adhesion. At this condition the tension 10,000 atmosphere to 31 atmosphere. It is the unavailable form of water. This condition mostly occurs at permanent wilting point stage or dry condition.

Hyperthermic: A soil temperature class with mean annual temperatures >22°C.

Hysteresis: A relationship between two variables that changes depending on the sequences or starting point. An example is the relationship between soil water content and water potential, for which different curves describe the relationship when a soil is gaining water or losing it.

Illuviation: The process of deposition in lower soil horizons of material eluviated (transported) from upper horizons.

Immobilization: The conversion of an element from the inorganic to the organic form in microbial tissues or in plant tissues, thus rendering the element not readily available to other organisms or to plants.

Immobile nutrient: not able to be moved from one part of the plant to another.

Infiltration: The downward entry of water into the soil or entry of water at the soil-air interface due to sorption and vertical flow of water through the soil profile.

Interveinal chlorosis: Yellowing in between leaf veins, yet veins remain green.

Ion exchange: The interchange between an ion in solution and another ion on the surface of any surface-active material such as clay or humus.

Irrigation efficiency: The ratio of the water actually consumed by crops on an irrigated area to the amount of water diverted from the source onto the area.

Isomorphous substitution: Substitution of one atom by another of similar size in a mineral without disrupting or changing the mineral structure.

Lacustrine deposits: The sediments deposited in fresh water lakes are known as 'lacustrine deposits'. The source of these sediments is erosion of lake banks and/or particles suspended in streams which get discharged into the lakes.

Land: resource that includes all natural resource coming from the earth (eg. Tree, animals, water and minerals), as well as physical space.

Latent heat: A property of a material. The amount of heat involved per unit mass when the material undergoes a change of phase (e.g., the heat of melting or the heat of vaporization).

Laterite: An iron-rich subsoil layer found in some highly weathered humid tropical soils that, when and allowed to dry, becomes very hard and will not soften the overlying layers, when rewetted.

Laterization: It is a pedogenic process of soil to removal of silica from soil and accumulation of sesquioxides.

Leaching: The removal in solution of soluble minerals and salts as water moves through the profile.

Ligand exchange: A class of surface reactions on minerals in which anions from solution take the place of some of the anions (ligands) normally bound to cations in the mineral.

Lime requirement: The mass of agricultural limestone, or the equivalent of other specified liming material, required to raise the pH of the soil to a desired value under field conditions.

Lithophile: Trace elements that have affinity for silicate minerals are termed as lithophiles. Among micronutrients, B, Zn, Fe, Mn and Ni are included in this group. The element listed within the parentheses shows secondary affinity.

Luxury consumption: The intake by a plant of an essential nutrient in amounts exceeding what it needs. For example, if potassium is abundant in the soil, alfalfa may take in more than it requires.

Marine deposits: Marine deposits are formed by the deposition of sediments carried by streams into the bodies of static water. The stream-carried sediments are eventually deposited in oceans, lakes, seas, and gulfs.

Mass flow: Movement of fluid in response to pressure. Movement of heat, gases, or solutes together with the flowing fluid in which they are contained.

Matric potential: That portion of the total soil water potential due to the attractive forces between water and soil solids as represented through adsorption and capillarity. It will never be positive and is negative until saturation wherein it is zero.

Metabolic quotient: A ratio of total microbial biomass carbon over total organic carbon. It typically ranges from 0.5 to 5% in soil.

Microbial biomass: Includes the smallest living organisms such as bacteria, fungus, protozoa, algae, actinomycetes, nematodes, and nonliving organisms: prions and viruses. Microbial biomass denotes a small portion (less than 5%) of soil organic carbon.

Microfauna: That part of the animal population which consists of individuals too small to be clearly distinguished without the use of a microscope. Includes protozoans and nematodes.

Microflora: That part of the plant population which consists of individuals too small to be clearly distinguished without the use of a microscope. Includes actinomycetes, algae, bacteria, and fungi.

Micronutrient: A chemical element necessary in only extremely small amounts (<50 mg/kg in the plant) for the growth of plants. Examples are B, Cl, Cu, Fe, Mn, and Zn.

Mineralization: The conversion of an element from an organic form to an inorganic state as a result of microbial decomposition.

Minimum tillage: A method of cultivation that reduces the number of machinery operations to the fewest required to create the proper soil condition for seed germination.

Moisture equivalent: The weight percentage of water retained by a previously saturated sample of soil 1 cm in thickness after it has been subjected to a centrifugal force of 1000 times gravity for 30 min.

Molarlity: A one-molar (M) solution contains one mole or one molecular weight in grams of a substance in each litre of the solution. The molar method of expressing concentration is useful because equal volumes of equimolar solutions contain equal number of molecules.

Mottling: The presence of more than one soil colour in a horizon. Mottling is often an indication of poor profile drainage but may be caused by the weathering of parent material.

Muck soil: A soil containing 20 to 50% organic matter or an organic soil in which the organic matter is well decomposed.

Mulch: Any material spread on the soil surface to protect soil from raindrops, sunshine, freezing, or evaporation.

Mutualism: The ecological interaction between two or more biological species where each species benefits.

Nitrification The biochemical oxidation of ammonium to nitrate, predominantly by autotrophic bacteria.

Nitrogen fixing: Microbes and their plant association bacteria are capable of reducing nitrogen (N_2) to ammonia (NH_3) with the help of an enzyme called nitrogenase. The process is known as 'Biological Nitrogen Fixation' (BNF).

Normality: Its can be defined as the number of equivalents of solute dissolved in 1 L of solution.

No-tillage: A farming system in which a crop is planted in the residue from a previous crop without soil tillage.

Nutrient accumulation: Storage of a nutrient in a particular part or portion of the plant is called nutrient accumulation. Different plant parts store nutrients differentially.

Nutrient content: Concentration of a nutrient or its amount per unit weight of a plant tissue is termed as nutrient content. Nutrient content is expressed in terms of percentage or ppm.

Nutrient uptake: Amount of nutrient taken up by the growing crops from either the soil or other sources, is called nutrient uptake. It is a product of plant matter produced and nutrient concentration in plant.

Octahedron: An eight-sided molecular unit. In mineralogy, a metal cation such as Al or Fe surrounded by six oxygen atoms or hydroxyl groups.

Organic fertilizer: Organic material that releases or supplies useful amounts of a plant nutrient when added to soil.

Osmotic potential: That portion of the total soil water potential due to the presence of solutes in soil water. It will generally be negative.

Oxidation ditch: An artificial open channel for partial digestion of liquid organic wastes in which the wastes are circulated and aerated by a mechanical device.

Oxidation: The addition of oxygen. More generally, the removal of electrons from an atom, ion, or molecule during a reaction. Oxidation may increase the positive charge of an element or compound.

Parasitism: A relationship between biological species, where one organism, the parasite, lives on or in another organism, the host, causing the host organism some harm, and the parasite organism is adapted structurally to this way of life.

Particle density: The weight per unit volume of the solid portion of soil. It is expressed as metric tons per cubic meter (Mg/m^3) or grams per cubic centimeter (g/cm^3).

Particulate organic matter: A microbially active fraction of soil organic matter consisting largely of fine particles of partially decomposed plant tissue.

Passive remote sensing: Remote sensing in which the energy being sensed is not supplied by the sensor. For example, any sensing that uses the sun's energy is passive remote sensing.

Peat: Unconsolidated soil material consisting largely of undecomposed, recognizable plant material or an organic soil containing more than 50% organic matter.

Ped: A natural unit of soil structure. Another term for aggregate.

Pedology: The science that deals with the formation, morphology, and classification of soil bodies as landscape components.

Pedon: The smallest volume that can be called a soil. It has three dimensions and lateral cross section is roughly hexagonal and ranges from 1 to 10 m^2 in size depending on the variability in the horizons.

Pedoturbation: Physical disturbance and mixing of soil horizons by such forces as burrowing animals (faunal pedoturbation) or frost churning (cryoturbation).

Percolation: The downward movement of water through soil. Especially, the downward flow of water in saturated or nearly saturated soil at hydraulic gradients of the order of 1.0 or less.

Permeability: The ability of a rock or soil to allow pore fluids and gases to pass through it. The is governed by the pore diameter and extent to which the pore spaces within the rock are connected

Persistence: The property of a material such as a pesticide or organic matter to resist microbial or chemical decomposition. The more persistent a chemical, the longer time required to decompose it.

pF: A logarithmic expression of soil suction or centimeter height of a water colume, whereby pF 2.5 is equivalent to a suction of 30 Kilopascals and represents field capacity for a prepared sample. pF 4.2 is equivalent to a suction of 1500 Kilopascals and represents permanent wilting point.

pH of soil: The negative logarithm of the hydrogen ion activity of a soil, expressed in moles/lit in aqueous solution and varying between 0 (extremely acid) to 14 (extremely alkaline).

Physiologically dry soil: The availability of water to plants is also influenced by dissolved salts. Consequently, a soil is considered to be dry when it is at moisture tension of 1500 kPa (15-bar) or more, or when salt concentration reaches the level that limits moisture availability to plants.

Phytoremediation: The use of plants to remediate (clean) soil by plant uptake and accumulation of soil contaminants.

Plastic limit: The moisture content at which a soil changes from semisolid to plastic.

Plasticity index: The numerical difference between the liquid limit and the plastic limit; the range of moisture content within which the soil remains plastic.

Plasticity: The capacity of organisms with the same genotype to vary in developmental pattern, in phenotype, or in behavior according to varying environmental conditions.

Plate count: A count of the number of colonies formed on a culture medium which has been inoculated with a small amount of organisms.

Platy: Soil aggregates that are horizontally elongated.

Podzolization: the removal of iron and aluminium often as complex with humus from the upper part and its deposition at some depth.

Polypedon: Two or more contiguous pedons, all of which are within the defined limits of a single soil series; commonly referred to as a soil individual.

Porosity: the percentage of soil volume occupied by pore space.

Porosity: The volume percentage of the total soil bulk not occupied by solid particles.

Potential acidity: The acidity that could potentially be formed if reduced sulfur compounds in a potential acid sulfate soil were to become oxidized.

Predation: It is the direct feeding of one organism on other.

Priming effect: The increased decomposition of relatively stable soil humus under the influence of much enhanced, generally biological, activity resulting from the addition of fresh organic materials to a soil.

Profile: A vertical section of the soil through all profile. So its horizons and extending into the parent material.

Protocooperation: In this kind of association both partners are mutually benefited like symbiosis.

Puddling: The the mechanical reduction in the apparent specific volume of soil.

Redox (Eh): Refers to oxidation-reduction, two processes that take place simultaneously. Oxidation is the loss of an electron by an atom, and reduction is the gain of an electron by an atom.

Redox potential: The electrical potential (measured in volts or millivolts) of a system due to the tendency of the substances in it to give up or acquire electrons.

Reduction: The gain of electrons, and therefore the loss of positive valence charge, by a substance. In some cases, a loss of oxygen or a gain of hydrogen is also involved.

Regolith: The unconsolidated mantle of weathered rock and soil material on the earth's surface; loose earth materials above solid rock.

Remote sensing: Identifying and observing objects at a distance without direct contact.

Residual acidity: Soil acidity that can be neutralized by lime or other alkaline materials but cannot be replaced by an unbuffered salt solution.

Resilience: The capacity of a soil to return to its original state after a disturbance.

Rhizosphere: That portion of the soil in the immediate vicinity of plant roots in which the abundance and composition of the microbial population are influenced by the presence of roots.

Rill erosion: An erosion process in which numerous small channels of only several centimeters in depth are formed; occurs mainly on recently cultivated soils.

Saline soil: A non sodic soil containing sufficient soluble salts to impair its productivity. The conductivity of a saturated extract is >4 dS/m, the exchangeable sodium adsorption ratio is less than about 13, and the pH is <8.5.

Saline-sodic soil: A soil containing sufficient ex-changeable sodium to interfere with the growth of most crop plants and containing appreciable quantities of soluble salts. The exchangeable sodium adsorption ratio is >13, the conductivity of the saturation extract is >4 dS/m (at 25°C), and the pH is usually 8.5 or less in the saturated soil.

Salinization: The process of making a soil saline; increasing the concentration of soluble salts.

Saltation: Soil particle movement in water or wind where particles skip or bounce along the stream bed or soil surface.

Savanna: The term savanna refers to tropical and subtropical vegetation formations with a predominantly continuous grass cover, occasionally interrupted by trees and shrubs.

Sedimentation: A technique for determining particle size distribution. Particles fall through water at a rate proportional to their size.

Sheet erosion: Water erosion that removes a uniform layer of soil from the land surface.

Shrinkage limit: The maximum water content at which a reduction in water content will not cause a decrease in the soil's volume.

Siderophile: Trace elements that have affinity for iron minerals are called siderophile. This group includes Fe, Cu, Mo and Ni.

Silicate:s: Those minerals in which silicon and oxygen are the major elemental constituents.

Slickensides: Subsoil structural features which develop as a result of two masses moving past each other, polishing and smoothing the surfaces. These are common in Vertosols.

Smectite: A family of expanding 2:1 (Si tetrahedral sheets to Al octahedral sheets) clay minerals with a high surface area and exchange capacity.

Sodium adsorption ratio: The concentration of Na divided by the square root of the sum of Ca and Mg concentrations, both expressed as molarities and measured in the saturation extract.

Soil acidification: is defined as the lowering of the soil pH because of the build-up of hydrogen and aluminium ions in the soil and the leaching of base cations such as calcium, magnesium, potassium and sodium.

Soil amendment: Any material, such as lime, gypsum, sawdust, or synthetic conditioner, that is worked into the soil to make it more amenable to plant growth.

Soil carbon sequestration: How carbon is stored in the soil; the process by which carbon dioxide from the atmosphere is converted to organic carbon by photosynthesis and the decomposition of that plant carbon into organic matter stored in the soil.

Soil colloid: Organic and inorganic matter with very small particle size and a correspondingly large surface area per unit of mass.

Soil compaction: is defined as the increase in density and a decline of macro-porosity in a soil that impairs the functions of both the top and subsoil, and impedes roots penetration and water and gaseous exchanges.

Soil conditioner: Any material added to a soil for the purpose of improving its physical condition.

Soil conservation: A combination of all management and land-use methods that safeguard the soil against depletion or deterioration caused by nature and/ or humans.

Soil delineation: The area enclosed by a boundary. Soil boundaries are shown on maps by lines.

Soil fertility: It is the ability of the soil to provide all essential plant nutrients in available forms and in a suitable balance.

Soil health: The capacity of a specific kind of soil to function, within natural or managed ecosystem boundaries, to sustain plant and animal productivity, maintain or enhance water and air quality, and support human health and habitation.

Soil microbiology: The branch of soil science concerned with soil inhabiting microorganisms, their functions, and activities.

Soil moisture regimes: The presence or absence of water in a soil at different times of a year.

Soil moisture tension: Soil moisture tension is the tenacity with which water is held in the soil. This pressure or tenacity is measured in terms of potential energy of water and is expressed in atmosphere or bars.

Soil monolith: A vertical section of a soil profile removed from the soil and mounted for display or study.

Soil organic carbon: Is related directly to soil organic matter. Soil organic matter consists of 50 to 58 percent soil organic carbon. A simple unit of 1.724 is used to convert total carbon into soil organic matter.

Soil productivity: The capacity of a soil for producing a specified plant or sequence of plants under a specified system of management. Productivity emphasizes the capacity of soil to produce crops and should be expressed in terms of yields.

Soil quality: The capacity of a soil to function within natural or managed ecosystem boundaries, to sustain plant and animal productivity, maintain or enhance water and air quality, and support human health and habitation.

Soil resilience: the capacity of soil to recover its function al and structural integrity after a disturbance.

Soil salinization: The increase in water soluble salts in soil which is responsible for increasing the osmotic pressure of the soil. In turn, this negatively affects plant growth because less water is made available to plants.

Soil sealing: The permanent covering of the soil surface with impermeable artificial materials such as asphalt and concrete.

Soil sodification: Is defined as an increase of the exchangeable sodium content of the soil, often accompanied by a loss of soil structure. In turn, it negatively affects soil suitability for crop growth.

Soil structure: The combination or arrangement of primary soil particles into secondary particles, units, or peds.

Soil texture: The relative proportions of the various soil separates (sand, silt and clay) in a soil.

Soil tilth: The physical condition of soil in relation to plant growth is related to size distribution of aggregates in soil.

Soil: The unconsolidated mineral materials on the immediate surface of the earth that serves as a natural medium for the growth of land plants.

Solum: The upper and most weathered part of the soil profile; the A, E, and B horizons.

Sorption: The removal from the soil solution of an ion or molecule by adsorption and absorption. This term is often used when the exact mechanism of removal is not known.

Specific gravity: The ratio of the weight or mass of a given volume of a substance to that of an equal volume of a reference substance. Water is the reference for liquids and solids.

Specific heat capacity: The amount of kinetic (heat) energy required to raise the temperature of 1 g of a substance (usually in reference to soil or soil components).

Specific surface: The solid particle surface area per unit mass or volume of the solid particles.

Splash erosion: The spattering of small soil particles caused by the impact of raindrops on very wet soils.

Subsoil: The B horizon of soil profile, not including the C horizon.

Surface tension: It is the total force acting in a solid-liquid-air system. The liquid surface has some properties of stretched elastic nature. This is due to the unequal forces of molecular attraction at the surface layer. This elasticity is known as surface tension.

Symbiosis: This is an association between two organisms where both the partners derive mutual benefit.

Tarai soils (Mollisols): Tarai soils are derived from the materials washed down by the erosion of mountains (alluvial origin). Relatively high moisture content for the greater part of the year results in luxuriant vegetation organic matter content is high Sandy loam to silty loam in texture.

Thermal diffusivity: It is defined as the ratio of thermal conductivity to heat capacity of a matter.

Tillage: The mechanical manipulation of soil for any purpose; but in agriculture it is usually restricted to the modifying of soil conditions for crop production.

Tilth: the physical condition of soil in relation to plant growth is related to size distribution of aggregates in soil.

Tracer nutrient: Radioisotope or a stable isotope of an element used for tracing its path in a system to study the mechanism of its interaction with the system is called a tracer element.

Unavailable water: Water that is present in the soil but cannot be taken up by plant roots because it is strongly adsorbed onto the surface of particles.

Unsaturated flow: The movement of water in a soil that is not filled to capacity with water.

Vadose zone: Unsaturated zone of soil above the groundwater, extending from the bottom of the capillary fringe all the way to the soil surface.

Vermiculite: An expanding 2:1 layer aluminosilicate clay mineral, in which one sheet of Al, O, and OH in octahedral coordination lies between two sheets of Si and Al, with O in a tetrahedral coordination.

Virgin soil: A soil that has not been significantly disturbed from its natural environment.

Void ratio: A measure of porosity. The ratio of void volume to soil bulk volume.

Water use efficiency: Dry matter or harvested portion of crop produced per unit of water consumed.

Waterlogging: Impeded drainage; or soil saturated with water but not necessarily with standing water.

Watershed: An area of land that collects and discharges water into a single stream or other outlet. Also called a catchment or drainage basin.

Weather: The state of the atmosphere in terms of such variables as temperature, cloudiness, precipitation, and radiation.

Weathering: All the physical, chemical and biological processes that cause the disintegration of rocks at or near the surface.

Wetting front: The advancing boundary between dry soil and wetted soil during infiltration.

Wilting point: It defines the amount of water remaining in the soil when a plant wilts and does not respond to added water (Measured at pF 4.2 and 15 atm bar).

Xerophytes: Plants that grow in or on extremely dry soils or soil materials.

Zero point of charge: The pH at which the net surface charge of the soil colloids is zero is known as the ZPC.

Zero tillage: can be defined as seeding in untilled conditions with the help of a specially designed fertiseedrill having inverted T type furrow openers. The sowing can be done at slightly below optimum moisture level to comparatively wet conditions.

Zonal soil: A soil type that is defined largely on the basis of climatic belts, where the influence of such factors as parent material is less evident than the influence of climatic development processes.

1
General

Historical Development of Soil Science/Agriculture

Francis Bacon (1561-1624 A.D.): Found the water as nutrient of plants.

Van Helmont (1577-1644): Give the ideas about the role of soil as a simple source of water and minerals for the plants.

Francis Bacon (1591-1624): suggested that the principle nourishment of plants was water and the main purpose of the soil was to keep plants erect and to protect from heat and cold.

G.R. Glanber (1604-1668 A.D.): Salt peter (KNO_3) as nutrient and not water.

Robert Boyle (1627-1691): an England scientist confirmed the findings of Van Helmont and proved that plant synthesis salts, spirits and oil etc from H_2O.

Jethrotull (1674-1741 A.D.): Fine soil particle as plant nutrient.

John Woodward (1699): the first documented scientist to successfully grow plants using hydroponics.

Priestly (1730-1799 A.D.): Established the essentiality of O_2 for the plant growth.

Anthur Young (1741-1820): an English agriculturist conducted pot experiment using Barley as a test crop under sand culture condition.

Theodore-de-Saussure (1767-1845): Found that plants absorb CO_2 from air and release O_2; soil supply N_2 and ash to plants.

Francis Home (1775 A.D.): Water, air, salts, fire and oil from the plant nutrients.

Thomas Jefferson (1793 A.D.): Developed the mould board plough.

J.B. Boussingault (1802-1882): French chemist conducted conduct field experiment. He is considered as father of field experiments.

Justus van Liebig (1804-1873 A.D.): A German chemist developed the concept called "Liebig's law of minimum".

J.B. Lawes and J. H. Gilbert (1843): established permanent manurial experiment at Rothemsted Agricultural experiment station at England.

J.T. Way (1856): Showed that nitrification is the conversion of ammonium into nitrate.

Hilgard (1892): Soil is more or less a loose and friable material in which plants, by means of their roots, find a foothold for nourishment as well as for other conditions of growth"

Whitney (1892): Soil is a nutrient bin which supplies all the nutrients required for plant growth.

N. M. Siburtzev (1890-1900): Write the first textbook of soil science, Siburtzev became the first professor of pedology.

Dokuchaiev (1900): Russian scientist - Father of soil science - Soil is a natural body composed of mineral and organic constituents, having a definite genesis and a distinct nature of its own.

F.T. Blackman's (1905): Theory of "Optima and Limiting Factors"

J.W. Leather (1906): was the first agricultural chemist at the Imperial Institute of Agricultural Research, Pusa, Bihar, India.

E.A. Mitsherlich (1909): proposed a theory of "Law of diminishing returns"

C.F. Marbut (1927): Classified soil based on zonality and emphasized the concept of pedalfers and pedocals.

Madam Scholasky (1932): The Earth, in which plants grow, composed of mineral and organic materials and living forms.

Joffe (1936): "Soil is a natural body of mineral and organic constituents differentiated into horizons - usually unconsolidated - of variable depth which differs among themselves as well as from the underlying parent material in morphology, physical makeup, chemical properties and composition and biological characteristics".

Macy (1936): Proposed a concept of "Critical Percentages of Plant Nutrients".

Hens Jenny (1941): Soil is a naturally occurring body that has been formed due to combined influence of climate and living organisms (active factors) acting on parent material as conditioned by relief over a period of time (passive factors).

Ruffin and Simonson (1968): Soil is a mixture of Earth's uppermost mantle of weathered rock and organic matter.

Buckman and Brady (1969): Soil is a dynamic natural body on the surface.

Discovery of Important Concepts/Terms in Soil Science

- A value: Fried & Dean
- Benchmark soil of India: Murthy
- Biogas: S.V. Desai & S.C. Biswas
- Capillary potential: Hillel (1972)
- Carrier hypothesis: Epstein (1953)
- Catena: Milne (1935)
- Cation exchange phenomenon: Thompsons (1850)
- Creeping death of the soil: Rama Rao
- Criteria of essentiality: Arnon (1954)
- Critical Percentages of Plant Nutrients: Macy (1936)
- Critical soil test level concept: Cate & Nelson
- Crop logging: H.F. Clement
- Desertification: Aubreville (1949)
- DRIS: Beaufils
- E value: Russell
- Electro ultra filtration theory: Nymith
- Electrochemical theory of Nutrient uptake: Landegadh
- Fertility gradient approach: B. Ramamoorthy
- Field plot technique: JB Boussingault
- Functional nutrients: Nicholas (1961)
- Ideotypes/Model plant: Donald in 1968
- Infiltrability: Hillal (1971)
- Integrated soil test approach: Colwell
- Inverse Yield–Nitrogen law: Wilcox (1929)
- L & Y value: Larsen
- Laterite: Buchanan (1807)
- Lattice Hole Theory: Page & Baver (1940)
- Law of diminishing returns: Mitscherlich.1909
- Leaching requirement: L.A. Richards
- Lime Potential: Aslying
- Lime Requirement: Bernstein and Francois
- Law of limiting Factor: Blakman

- Law of Minimum: Von Liebig in 1840
- Law of restitution: Justus von liebig (1840)
- Malthusian Theory: Thomas Malthus (1898)
- Mixed cropping: La flitze (1928)
- Moisture equivalent: Briggs& Melane (1907)
- Nutrient index: Parker
- Nutrient mobility concept: Bray
- Nutrient sufficiency: Baule
- Optima and Limiting Factors: F.T. Blackman's (1905)
- pE: Silen
- Pedon: Simonson
- Permanent wilting point: Briggs & Shantz
- pF: Schofield
- Q/I: Beckett (1964)
- Rating chart of nutrient: Mohar
- Rizospheres: Hiltner (1904)
- Salt index: Puri
- Sodium Absorption Ratio: Richards
- Soil zonelity: Dokuchave
- Stability index of mineral: Jakson
- Target Yield: Ramamorthy (1967)
- Terminal Velocity Low: G G Stokes
- Ultra micronutrient (CO): Nicholas
- Water harvesting: Myors
- Water Potential: Buckingham (1907)
- Zonality concept: Dokuchaiev

Pioneers in Soil Scinece

- Father of Soil Science – **Dakuchaev**
- Father of Indian Soil Science & Agricultural Chemistry- **J. W. Leather**
- Father of Soil Testing – **Troug**
- Father of Soil Physics – **Hillel**
- Father of Zero Tillage, Weeds and Seed drill – **Jethrotull**

- Father of Modern Tillage - **Jethrotull**
- Father of Microbiology – **Louis Pasteur**
- Father of Soil Microbiology - **Sergei Winogradsky**
- Father of American Soil Microbiology- **Waksman**
- Father of Agricultural Chemistry – **Justus Von Liebig.**
- Father of Pedology - **Dakuchaev**
- Father of Natural Farming – **Masanuba Fukuaka**

International and National Journal of Soil Science

Name of Journal	Publisher
Acta Agriculturae Scandinavica, Section B-Soil and Plant Science	Taylor & Francis, United Kingdom
Applied Soil Ecology	Elsevier BV Netherlands
Archives of Agronomy and Soil Science	Taylor & Francis, United Kingdom
Biology and Fertility of Soils	Springer, Germany
Canadian Journal of Soil Science	Agricultural Institute of Canada, Canada
Clean - Soil, Air, Water	Wiley, Germany
Communications in Soil Science and Plant Analysis	Taylor & Francis, United Kingdom
European Journal of Soil Biology	Elsevier BV, Netherlands
European Journal of Soil Science (Journal of Soil Science)	Wiley, Germany
Indian Journal of Soil Conservation	Indian Associations of Soil & Water Conservationists, India
Journal of Plant Nutrition and Soil Science	Wiley, Germany
Journal of Soil and Water Conservation	Soil and Water Conservation Society, USA
Journal of Soil Biology and Ecology	Indian Society of Soil Biology and Ecology, India
Journal of Soil Salinity and Water Quality	CSSRI, Karnal, India
Journal of Soil Science and Plant Nutrition	Springer Nature, Switzerland
Journal of Soils and Sediments	Springer Nature, Switzerland
Journal of the Indian Society of Soil Science	Indian Society of Soil Science, India
Plant and Soil	Springer Nature, Switzerland
Plant Soil and Environment	Czech Academy of Agricultural Sciences, Czechoslovak, Koruna
Soil and Environment	Soil Science Society of Pakistan
Soil and Sediment Contamination	Taylor & Francis, United Kingdom
Soil and Tillage Research	Elsevier BV, Netherland
Soil Biology and Biochemistry	Elsevier BV, Netherlands
Soil Research (Australian Journal of Soil Research)	CSIRO Publishing, Australia

Name of Journal	Publisher
Soil Science	Lippincott Williams & Wilkins, United States
Soil Science and Plant Nutrition	Taylor & Francis, United Kingdom
Soil Science Society of America Journal	Soil Science Society of America
Soil Use and Management	Wiley, Germany
Water, Air and Soil Pollution	Springer Nature, Switzerland

Soil Science and Allied Research Institutes in India

- Indian Institute of Soil Science (IISS), Bhopal
- Central Soil Salinity Research Institute (CSSRI), Karnal
- National Bureau of Soil Survey and Land Use Planning (NBSSLUP), Nagpur
- Indian Institute of Soil and Water Conservation (IISWC), Dehradun
- Central Research Institute for Dryland Agriculture (CRIDA), Hyderabad
- Central Coastal Agricultural Research Institute (CCARI), Goa
- Central Arid Zone Research Institute (CAZRI), Jodhpur
- Central Agroforestry Research Institute (CAFRI), Jhansi
- Indian Institute for Farming System Research (IIFSR), Modipuram
- Indian Institute of Water Management (IIWM), Bhubaneswar
- National Organic Farming Research Institute (NOFRI), Tadong, Sikkim
- National Institute of Abiotic Stress Management (NIASM), Baramati
- National Research Centre for Integrated Farming (NRCIF), Motihari
- ICAR Research Complex for Eastern Region, Patna
- ICAR Research Complex for NEH Region, Barapani

Soil Nutrient Chart

Parameters	Low	Medium	High
pH (1:2.5)	<6.5 (Acidic)	6.5-7.5 (Neutral)	>7.5 (Alkaline)
EC (dS/m)	<1.0	1-2	>2.0
O.C. (g/kg)	<4.0	4-8	>8.0
$CaCO_3$ (%)	<3.0	3-8	>8.0
N (kg/ha)	<280	280-560	>560
P (kg/ha)	<14	14-28	>28
K (kg/ha)	<150	150-250	>250
S (mg/kg)	<10.0	10-20	>20.0
Zn (mg/kg)	<0.60	0.6-1.80	>1.80
Fe (mg/kg)	<4.50	4.50-18.0	>18.0
Cu (mg/kg)	<0.20	0.20-0.80	>0.80
Mn (mg/kg)	<2.0	2.0-8.0	>8.0
B (mg/kg)	<0.50	0.50-1.0	>1.0
Mo (mg/kg)	<0.10	0.10-0.40	>0.40

Electrical Conductivity Chart

EC (1: 2 soil-water) (dS/m)	Salt concentration (%)	Category
Below 0.8	Below 0.05	Normal
0.8 -1.6	0.05-0.15	Critical for salt sensitive crops
1.6 -2.5	0.15-0.20	Critical for salt tolerant crops
Above 2.5	0.20-0.25	Injurious to all crops

Category	pHs	ECe	ESP
Normal	6.5-7.5	<4.0	<15.0
Saline	< 8.5	>4.0	<15.0
Saline Alkali	> 8.5	>4.0	>15.0
Alkali	> 8.5	<4.0	>15.0

2
Soil Fertility

1. White alkali soils also know as

   a. Acid soil
   b. Saline soil
   c. Saline sodic soil
   d. Sodic soil

2. The most deficient micro-nutrient in Indian soil is

   a. Zn
   b. Fe
   c. Mn
   d. B

3. The pop in groundnut is due to deficiency of

   a. Ca
   b. N
   c. P
   d. K

4. Black cotton soils contain clay in the range of

   a. 10 - 20%
   b. 30 - 80%
   c. > 90%
   d. > 80%

5. Grass tetany in forage crops is caused by deficiency of

   a. Mg
   b. S
   c. Ca
   d. Mn

6. The quantity of urea needed for applying 120 kg nitrogen to rice will be

   a. 240 kg
   b. 250 kg
   c. 260 kg
   d. 270 kg

7. What is the optimum pH for the availability of most of the plant nutrients ?

   a. 5.0-6.0
   b. 6.0-7.0
   c. 6.5-7.5
   d. 6.0-8.0

8. Volumetric composition of porous space in soil is

a. 25% b. 65%

c. 50% d. 5%

9. Sulphur content in single super phosphate is

a. 8% b. 12%

c. 16% d. 18%

10. How much of the plant dry weight is constituted by C, H and O

a. 50% b. 70%

c. 30% d. >90%

11. Which types of soil structure is most important from crop growth point of view?

a. Platy b. Prismatic

c. Crumby d. Granular

12. pH of most of the mineral soil lying in the range

a. 0-14 b. 1.5-13.5

c. 6-8 d. 3-11

13. Choose the incorrect statement

a. Excess copper may enhance the utilization of molybdenum

b. Copper deficiency hinders cell division in plants

c. Excess potassium enhances manganese uptake

d. Heavy nitrogen fertilization intensifies copper deficiency

14. Phosphorus is mostly taken by plants in the form of

a. H_3PO_4 b. PO_4^{3-}

c. $H_2PO_4^-$ d. P^{5+}

15. Which of the following is correct ?

a. $P_2O_5 \times 0.44 = P$ b. $P_2O_5 \times 2.29 = P$

c. $P_2O_5 \times 0.83 = P$ d. $P_2O_5 \times 1.20 = P$

16. One centimetre water in a hectare area is equal to

a. 3530 foot^3 b. 100 m^3

c. 100000 litre d. All of the above

17. Puddling is needed in paddy cultivation to minimise

a. Evaporation loss b. Evapo-transpiration loss

c. Percolation loss d. Conveyance loss

18. The process of biological conversion of ammonia into nitrate is known as

a. Ammonification b. Nitrification

c. Humification d. Denitrification

19. The characteristics of saline alkali soil is

a. EC>4 dS/m, pH>8.5, ESP<15 b. EC>4 dS/m, pH<8.5, ESP>15

c. EC>4 dS/m, pH>8.5, ESP >15 d. EC>4 dS/m, pH<8.5, ESP<15

20. Saline soils posses

a. Very high salt concentration b. Very high sodium concentration

c. Both (a) and (b) d. None of above

21. The soils are strongly calcareous if they contain

a. Less than 5% $CaCO_3$ b. Greater than 10% $CaCO_3$

c. 5 to 10% $CaCO_3$ d. All above

22. National Project on Management of Soil Health and Fertility scheme was introduced in

a. 2004-05 b. 2008-09

c. 2011-12 d. 2015-16

23. Soils get dispersed if exchangeable sodium

a. Goes beyond 15% b. Less than 15%

c. Up to 15% d. All above

24. In which soil, zinc availability to crop is higher ?
    a. Acidic soil
    b. Alkaline soil
    c. Calcareous soil
    d. Waterlogged soil
25. Gypsum requirement of soil is based
    a. On ESP of soil
    b. On CEC of soil
    c. Both (a) & (b)
    d. None of above
26. The readily available form of K in soil ranges from
    a. 0- 2.0 %
    b. 1- 10 %
    c. 30 - 40 %
    d. 90 -98%
27. Zinc (Zn) deficiency was first diagnosed in rice on
    a. Deseret soil
    b. Black cotton soils
    c. Calcareous soils
    d. Laterite soil
28. Availability of which micronutrient increases with increase in soil pH
    a. B
    b. Fe
    c. Zn
    d. Mo
29. Inherent capacity of soil to supply nutrients in balanced form is called
    a. Soil productivity
    b. Soil fertility
    c. Fertility index
    d. Productivity index
30. Deficiency of K first appears on which part of the plants
    a. Older leaves
    b. Middle leaves
    c. Younger leaves
    d. Terminal full leaves
31. Which nutrients is known as immobile nutrient element
    a. Potassium
    b. Sulfur
    c. Calcium
    d. Zinc
32. Which nutrient elements imparts the role of indirect drought resistance
    a. Potassium
    b. Calcium
    c. Magnesium
    d. Sulfur

33. Ammonium sulphate is advisable for which of the following crop

a. Wheat
b. Maize
c. Tea
d. Cotton

34. Well rotted FYM contain N

a. 0.5%
b. 1.0%
c. 1.5%
d. 2.0%

35. 'Mg' and 'S' are termed as

a. Primary nutrients
b. Ultra-micronutrient
c. Micronutrient
d. Secondary nutrient

36. Chlorosis refers to

a. Burning of leaf tissue
b. Death of leaf tissue
c. Yellowing of leaf tissue
d. Development of reddish brown colour on leaf

37. Which one of the following is an example of natural chelate

a. EDTA
b. DTPA
c. Citric acid
d. EDDHA

38. Which micronutrient is essential for the microbial assimilation of nitrogen

a. Mo
b. B
c. Zn
d. Cu

39. Khaira disease of rice is due to deficiency of

a. Zinc
b. Iron
c. Nitrogen
d. Calcium

40. Which plant element accounts for at least one half of the total number of ions absorbed

a. N
b. P
c. K
d. Ca

41. National Centre of Organic farming was established in year ______ with an objective to promote organic farming in the country.

    a. 2004 b. 2008

    c. 2010 d. 2012

42. Which one of the following elements is a micronutrient.

    a. Boron b. Sulphur

    c. Calcium d. Magnesium

43. Which one is dominating salt in alkali soil ?

    a. $CaCl_2$ b. $MgSO_4$

    c. $NaHCO_3$ d. Na_2CO_3

44. pH of saline soils is

    a. 8.5 b. 9.5

    c. 10.5 d. 11.5

45. Sodic soils can be made more productive by adding

    a. Boron b. Calcium

    c. Phosphorus d. Sodium

46. Which one of following is commonly used for naval disinfection

    a. Distilled water b. Tincture of iodine

    c. Sodium chloride d. Iodex

47. Targeted yield concept was given by

    a. Marshall b. Libby *et al.*

    c. Sorenson d. Ramamoorthy *et al.*

48. Whiptail in Brassica is due to the deficiency of

    a. Calcium b. Molybdenum

    c. Manganese d. Zinc

49. Which fertilizer contains nutrients other than primary nutrients

    a. Urea b. Diammonium Phosphate

    c. Ammonium Sulphate d. Ammonium Nitrate

50. High salt tolerant crop is

    a. Rice b. Pea

    c. Gram d. Sunhemp

51. Akiochi disease of rice commonly found in ________

    a. Sodic soil b. Acidic soil

    c. Saline soil d. Alkaline soil

52. An ordinary rain will generally be

    a. Slightly acidic b. Slightly alkaline

    c. Highly acidic d. Highly alkaline

53. Khaira disease of rice can be controlled by spraying

    a. Copper sulphate b. Borax

    c. Manganese sulphate d. Zinc sulphate

54. Sulphur content in ammonium sulphate is

    a. 12 % b. 20 %

    c. 24 % d. 30 %

55. Indicator plant of copper is

    a. Maize b. Wheat

    c. Sorghum d. Potato

56. Which element has a role in energy transformation and metabolic processes in plants

    a. N b. P

    c. Mo d. S

57. Aspergillus niger is used to test the status of which element in the soil

    a. N b. P

    c. K d. Fe

58. What is the estimated extent of salt affected soils in India?

    a. 5.42 mha b. 7.42 mha

    c. 11.42 mha d. 17.42 mha

59. When a fertilizer is used for supplying a single nutrient, it is called as

a. Straight fertilizers
b. Binary fertilizers
c. Mixed fertilizers
d. Solid fertilizers

60. RSC safe for irrigation water is

a. <1.25
b. > 2.5
c. 1.25- 2.5
d. 5.0

61. Soils having 200 kg available K_2O/ha will be classed as

a. Low
b. Medium
c. High
d. None of these

62. The majority plant nutrients are available at pH

a. 8.5-11.0
b. 6.5- 7.0
c. 4.0-6.0
d. 8.5

63. Plant samples are generally dried at

a. 105°C
b. 60-70°C
c. 100°C
d. 25°C

64. Iron & Aluminium ions are predominant in

a. Acid soil
b. Saline soil
c. Calcareous soil
d. Sodic soil.

65. Who proposed the term 'functional or metabolic nutrient' in plant nutrition?

a. D.J. Nicholas
b. Arnon
c. Stout
d. Arnon and Stout

66. Percentage of Nitrogen in a matter calculated by

a. Organic carbon x 1.3
b. Organic carbon
c. Organic matter - organic carbon
d. Percentage organic carbon/10 (%C/10)

67. Which of the following method gives an index of soil phosphorus under biological method of soil fertility evaluation

a. Aspergillas niger method
b. Mitcherlic pot culture method
c. Jenny's pot culture method
d. Cunning ghamela plaque method

68. Choose the correct statement in given option

    a. Inorganic fertilizer is urea and calcium cynamide

    b. Permission limit of buret content in urea more than 1.5%

    c. Rock phosphate used in high pH soil.

    d. The protein, content of nitrogen is 16%

69. The phosphorus bearing compounds in soils mean

    a. Phytins
    b. ATP
    c. Both a and b
    d. Proteins

70. Which of the following element is tested through aspergillas- method of soil fertility evaluations

    a. Nitrogen
    b. Potassium
    c. Phosphorus
    d. Sulphur

71. Which of the following scientists have established nickel as a essential element for pant growth during 1987

    a. P.R. Brown, R.N. Welsh
    b. K Warrington E.E. Carr
    c. Springel
    d. Ahinad and Evan

72. Identify the following micronutrient element involved in reproductive growth of plant.

    a. Boron
    b. Iron
    c. Zinc
    d. Nickel

73. Which of the following beneficial element is required for microorganisms fixing atmospheric nitrogen

    a. Sodium
    b. Vanadium
    c. Cobalt
    d. Silicon

74. The concept of leaching requirement was developed by

    a. W.R. Gardner
    b. L.A. Richards
    c. L.D. Baver
    d. D. Kirkham

75. The amount of 'P' in soil and soil solution that can exchange with ortho phosphate in addition to soil system is termed as

    a. L- value
    b. A-value
    c. E-value
    d. None of the above.

76. Which of the following range of watersoluble `K' content in soil

    a. 0.65-4.0 mg/100 g of soil
    b. 70.5-162.4 mg/100g of soil
    c. 6.5-37.2 mg/100 g of soil
    d. None of these

77. Which one of the main process of nitrate loss occur in wet soil after fertilizer application

    a. Volatilization
    b. Adsorption
    c. Leaching
    d. Fixation

78. The nutrient index value : Area with nutrient index value is > 2.3, then

    a. High
    b. Low
    c. Medium
    d. Low medium

79. Mehlichs cunmiughamella plague method is used to test

    a. N
    b. P
    c. Mn
    d. Zn

80. Sugarbeet is an indicator plant for

    a. Phosphorus
    b. Boron
    c. Magnesium
    d. Sodium

81. The Sulphur (S) content in pure gypsum is

    a. 21%
    b. 23%
    c. 26%
    d. 16%

82. Reproduction in plants is associated with deficiency of

    a. N
    b. Ca
    c. P
    d. Mg

83. Plants survive in saline soil is known as

    a. Oxlophyte
    b. Tharmophyte
    c. Holophyte
    d. Zerophyte

84. The conversion factor to convert % P to P_2O_5 is

a. 1.29
b. 2.90
c. 2.29
d. 0.40

85. Hill soil are generally

a. Neutral
b. Acidic
c. Saline
d. Alkaline

86. Farm Yard Manure (FYM) have N, P_2O_5 and K_2O approximately

a. 1.0, 0.6 and 1.0%
b. 2.0, 1.0 and 0.5%
c. 0.5, 0.2 and 0.5%
d. 3.0. 2.0 and 1.0%

87. A mature and dry plant tissue contain about proteins (%)

a. 20 - 50
b. 1-15
c. 1 - 8
d. 10 - 30

88.The pH range of calcareous soils is mostly

a. Less than 6.5
b. 6.5 to 7.5
c. 7.5 to 8.5
d. More than 8.5

89. Browning disease of rice is a

a. Physiological disease
b. Fungal disease
c. Bacterial disease
d. Viral disease

90. Pick the beneficial element of plant growth

a. Fe
b. Si
c. Co
d. Mo

91. Phosphorus absorbed by plant roots in acidic condition

a. PO_4^{++}
b. $P_2O_3^{+}$
c. $H_2PO_4^{-}$
d. P^{++}

92. Nutrient deficiency of which nutrient causes chlorosis and stunted growth

a. Na
b. Mo
c. Cl
d. N

93. Which fertilizer shows luxury consumption in plant

a. N
b. P
c. K
d. Zn

94. Which one of the following is the correct quantity of zinc sulphate required to prepare 100 litres of 200 ppm solution for foliar spray sulphate required (in grams)

a. 20
b. 100
c. 200
d. 2000

95. Molybdenosis, a disease in cattle, is caused by an imbalance of Mo and Cu in the diet, when Mo content of the forage is greater than

a. 5 ppm
b. 10 ppm
c. 15 ppm
d. 20 ppm

96. Which of the following nutrient impart resistance to plants against fungal and bacterial diseases

a. N
b. P
c. K
d. Ca

97. High analysis fertilizers are those

a. Which require high cost of analysis
b. Which require high level of analysis procedure
c. Which contain, high percentage of nutrient element
d. Whose analysis gives higher percentage of nutrient than actual content

98. Which one of the following has the highest iron content (%)

a. Ferrous sulphate
b. Ferrous ammonium phosphate
c. Ferrous oxide
d. Ferrous ammonium sulphate

99. The (KCl) murate of potash is not suitable for tobacco and potato because of

a. Cl toxicity
b. K toxicity
c. Costly fertilizer
d. Not readily available

100. The pH range at which Fe solubility occurs is

a. Less than 6.5
b. 7.5 to 8.5
c. 8.5 to 9.5
d. More than 9.5

101. How much % P_2O_5 is contained in triple super phosphate

a. 25%
b. 48%
c. 40%
d. 16%

102. In humid region generally

a. Acid soils are formed
b. Alkali Soils are formed
c. No soil formation takes place
d. All of these

103. Lime requirement is calculated for

a. Saline soil
b. Alkaline soil
c. Sodic soil
d. Acidic soil

104. Deficiency symptom of potassium is first appeared in

a. Upper/new leaf
b. Lower/older leaf
c. In top buds
d. All of these

105. Which one of the following will perform very poor as a source of N fertilizer under waterlogged rice field

a. Ammonium nitrate
b. Urea
c. Ammonium chloride
d. Ammonium sulphate

106. Improper development of wheat inflorescence is due to deficiency of :

a. Iron
b. Zinc
c. Phosphorus
d. Boron

107. The availability of molybdenum is more at

a. pH-5.0
b. pH-7.5
c. pH-9.0
d. pH-7.0

108. Which of the following mechanism is put under active transport of ions from soil to plant roots

a. Ion exchange
b. Protein synthesis
c. Carrier hypothesis
d. Content exchange.

109. Excesses of phosphorus in soil may cause deficiency of which micro-nutrients

a. Cu & Ni
b. Mn & B
c. Fe & Zn
d. Ni & Cd

110. Which of the following element deficiency in calcareous soil causes "Lime induced chlorosis" in plant

a. Mn
b. Cu
c. Zn
d. Fe

111. Which of the following worms are to be used as branding worms in Vermicomposting.

a. Lamricus ruboolus
b. Eisenia foetida
c. Allolobophora Caligonosa
d. None of these.

112. Which of the following organic manures contains organic substances which act as chelating compound

a. F.Y.M
b. Vermicompost
c. Humus
d. Compost

113. Which of the following 'P' fertilizers form the cakes under moist conditions

a. Diammonium phosphate
b. Single super phosphate
c. Triammonium phosphate
d. None of these

114. Carbon (C), Hydrogen (H) and Oxygen (O) constitute about in plants

a. 95 - 96%
b. 80 - 85%
c. 90 -95%
d. 85 -90%

115. When the plant absorb of nutrient in higher quantity but not effect on yield is know as

a. Hidden hunger
b. Luxury consumption
c. Intensity factor
d. None of these

116. Criteria of essential plant nutrients was given by

a. Bray
b. Beaufils
c. Jackson
d. Arnon and stout

117. If a soil sample content 6.0 g/kg of organic carbon, what will be organic matter in soil sample?

a. 0.6 %
b. 6.0 %
c. 1.0 %
d. 0.1 %

118. How many percentage of Sulphur found in gypsum ($CaSO_4.2H_2O$)

a. 19
b. 23
c. 26
d. 29

119. What will be the Leaching Requirement (L.R.) of the given irrigation water having EC 4 dS/m when EC of drainage water is 10 dS/m

a. 20.5%
b. 35.4%
c. 40.0%
d. 40.3%

120. BGA was discovered as a nitrogen fixer in paddy field (1939) by

a. P K dey
b. Beijerinck
c. Fred
d. Hiltner

121. Which of the following micronutrient is require in nitrogenases enzymes

a. Vanadium and cobalt
b. Molybdenum and Iron
c. Rubidium
d. All of these

122. Mn deficiency causes

a. Speckled yellow of sugar beet
b. Grey speck of oat and Marsh spot of pea
c. Pahala blight of sugarcane and frenching of tung trees
d. All of these

123. Rock phosphate most suitable fertilizer in conditions

a. Cat soil
b. Alkaline soil
c. Saline soil
d. None of these

124. Toxicity is probably the important growth limiting factor in acid soils, particularly when pH < 5.0 to 5.5

a. Fe
b. H^+
c. Al
d. Mn

125.CEC values for organic soils lies in range of

a. 3 - 5 meq/100 g
b. 10 - 20 meq/100 g
c. 20 - 50 meq/100 g
d. 50 - 100 meq/100 g

126. Which of the following amino acid accumulates under water stress condition in plants

a. Proline
b. Tyrosine
c. Glycine
d. All of the above

127. Highly hygroscopic fertilizer is

a. Urea
b. Ammonium sulphate
c. Ammonium nitrate
d. Both (a) and (c)

128. 13th most abundant element in the earth crust is

a. Calcium
b. Sulphur
c. Magnesium
d. Silicon

129.For 92 kg nitrogen, how much quantity of urea is required

a. 120 kg
b. 200 kg
c. 46 kg
d. 184 kg

130. Maximum saline and alkaline soils are found in

a. Uttar Pradesh
b. Punjab
c. Rajasthan
d. Gujarat

131. Lest amount of water-soluble phosphorus is found in

a. DAP
b. Rock phosphate
c. Single super phosphate
d. Tricalcium phosphate

132. Which of the nitrogenous fertiliser under amide group?

a. Ammonium sulfate
b. CAN
c. Sodium nitrate
d. Urea

133. Which types of fertilizer in India is fully dependant on imports

a. N-fertilizer
b. P-fertilizer
c. K-fertilizer
d. P and K fertilizer

134. Mostly plant takes the nitrogen in the form of

a. Nitrite (NO_2)
b. Nitrate (NO_3)
c. Ammonia (NH_4)
d. All of these forms

135. Intracellular deposits of which element is known as plant opals ?

a. Sodium
b. Silicon
c. Nickel
d. Vanadium

136. Which fertilizer is organic fertilizer

a. Urea
b. Calcium cyanide
c. CAN
d. Both a and b

137. The capacity of a soil resist appreciable change in pH value is called

a. Cation exchange capacity
b. Buffering capacity
c. Anion exchange capacity
d. Base saturation percentage

138. Fe concentration used for foliar spray is

a. 0.2%
b. 0.4%
c. 0.05 %
d. 0.6%

139. Equivalent acidity of anhydrous ammonia is

a. 148
b. 128
c. 110
d. 93

140. Essential element which is principle component of nitrate reductase is

a. Magnesium
b. Molybdenum
c. Boron
d. Zinc

141. Lime recommendation is made on the basis of a ...... inch furrow slice

a. 2
b. 4
c. 6
d. 15

142. The basic principle of conservation agriculture:

a. No tillage
b. Minimum disturbance of soil
c. >30% plant residue on surface
d. All of these

143. Which micronutrient is fixed on the same sites of the soil colloids as is phosphate

a. Boron
b. Copper
c. Zinc
d. Molybdenum

144. The amount of nitrogen in the arable soils in the form of NH_4^+ and NO_3 expressed as percentage of total nitrogen is

a. <10%
b. 1-2%
c. >5%
d. 5-10%

145. Which types of cropping system are more removal of zinc micronutrient in soil

a. Rice - Wheat
b. Maize - Wheat
c. Soybean - Wheat
d. Pigeonpea - Wheat

146. Which micronutrient is not absorb by plant in the metal ionic form from soil solution

a. Boron
b. Copper
c. Zinc
d. Iron

147. Internal cork disease in apple is caused by the deficiency of

a. Mn
b. B
c. Ca
d. Fe

148. Soils having grey brown upper crust and are hard on walking are known as

a. Sodic soils soil
b. Normal soils soil
c. Saline soil
d. Saline alkali soil

149. The surface run off during rains leads to decreased soil fertility mainly in terms of

a. Nitrate losses
b. Micronutrient losses
c. Phosphate losses
d. Nitrogen and phosphate

150. Mitscherlich related the potential yield of a crop to the amount of a given nutrient with the following equation: Log (A-Y) = log A-Cb, In the equation 'b' denotes

a. Proportionality constant
b. Potential yield
c. Amount of given form of nutrient present in the soil
d. Growth factor specific to a crop when all other factors are not limiting

151. Which is the most widely prevalent clay mineral in Indo-Gangetic alluvial soil

a. Kaolinite
b. Montmorillonite
c. Vermiculite
d. Illite

152. The fertilizer that may cause maximum phytotoxicity when not placed at appropriate depth in the soil is

a. Urea phosphate
b. Urea
c. Ammonium nitrate
d. Ammonium chloride

153. Non essential but useful elements for plants would be

a. Carbon and Vanadium
b. Vanadium and Chlorine
c. Vanadium and Sodium
d. Sodium and Chlorine

154. Rock phosphate is used in

a. Neutral soil
b. Alkaline soil
c. Acidic soil
d. saline soil

155. Crinkle leaf of cotton is caused by

a. Fe toxicity
b. Cu toxicity
c. Mn toxicity
d. Zn toxicity

156. Which of the following soils have poor physical condition ?

a. Saline
b. Sodic
c. Saline - alkali
d. None

157. Near the canal banks soils become in nature

a. Neutral soil
b. Alkaline soil
c. Acidic soil
d. Well drained soil

158. Terai soils are rich in

a. Mineral matter
b. Organic matter
c. Phosphate ion
d. Soluble sat

159. The nutrient availability concept that does not include growing plant system is

a. E-value
b. L-value
c. A-value
d. None of these

160. When Fe contents in plants falls below or equal to ..... ppm, deficiency is likely to occur.

a. 20
b. 40
c. 50
d. 100

161. Liebig's law of minimum for growth reaction is

a. Linear
b. Paraboli
c. Curvilinear
d. Hyperbolic

162. Cell wall thickening in root apical meristem occurs due to deficiency of

a. Zn
b. Mo
c. B
d. Mn

163. Amount of NPK in poultry manure : (N: P: K) is

a. 0.5: 0.3 :0.5 %
b. 3.0 :1.0 :2.0 %
c. 3.0 :2.5 :1.5 %
d. 0.7 :1.5 :1.2 %

164. Solubility of gypsum is

a. 5%
b. 0.25%
c. 50%
d. 0.10%

165.......... index is used in the process of land evaluation for land use planning

a. Erodibility
b. Storie
c. Productivity
d. Fertility

166. The availability of nutrient from soil reserves depends on

a. Quantity factor
b. Intensity factor
c. Both A and B
d. Buffering

167. How much of muriate of potash (KCl-60% K_2O) is required for supplying 50 kg K_2O/ha

a. 53.3 kg
b. 63.3 kg
c. 73.3 kg
d. 83.3 kg

168. The phosphatic fertilizer most suitable for calcareous soil is

a. Dicalcium phosphate
b. Diammonium phosphate
c. Rock phosphate
d. Single super phosphate

169. Which micronutrient is usually deficient in calcareous soils

a. Mn
b. Mo
c. Cu
d. B

170. C:N ratio of FYM is

a. 20:1
b. 40:1
c. 50:1
d. 80:1

171. Whiptail in Cauliflower is a characteristic indication of....... deficiency

a. Molybdenum
b. Boron
c. Calcium
d. Iron

172. Most suitable soil for Potato and Groundnut

a. Sandy loam
b. Clay loam
c. Sandy
d. Clay

173. The organic manure which has the narrowest C:N ratio is

a. Biogas slurry
b. Farm yard manure
c. Vermicompost
d. Groundnut cake

174. What is chemical formula of urea

a. NHCONH
b. $NHCONH_2$
c. NH_2CONH_2
d. NH_2CONH_3

175. Dolomite is used in fertilizer mixture as

a. Filler
b. Neutralizer
c. Conditioner
d. Stabilizer

176. Volumetric composition of organic matter in soil is

a. 25%
b. 45%
c. 50%
d. 5%

177. Sulphur content (%) in potassium sulphate is

a. 50
b. 24
c. 12
d. 17

178. Which soils are also called cat-clays'

a. Acid sulfate soils
b. Desert soils
c. Lateritic soils
d. Podsols

179. Which gas is emitted due to rice cultivation

a. CO
b. CO_2
c. CH_4
d. NH_4

180. Which one of the following element is considered to be an immobile element in the plant

a. Ca
b. Mg
c. N
d. P

181. Among the four which is an essential micronutrients required for plant growth

a. Aluminium
b. Cobalt
c. Nickel
d. Copper

182. In sugarcane, which leaf is most appropriate for total tissue test

a. The second leaf from the top of the plant
b. The third leaf from the top of the plant
c. Fully mature leaves midway between the younger centre leaves and the oldest leaf whorl on the outside
d. Fourth fully developed leaf from the top

183. Which part of cotton plant is used for tissue test ?

a. All the above ground portion
b. The youngest fully mature leaves of the main stem

c. Fourth fully developed leaf from the top

d. The four upper leaf blades

184. Which element is indirectly related with drought resistance

a. N
b. K
c. P
d. Zn

185. The amount of nitrogen fixation capacity of soybean in kg/ha per year is

a. 49 - 130
b. 30 - 60
c. 15 - 30
d. 12-15

186. How many elements are essential for the growth and development of plants

a. 14
b. 15
c. 16
d. 17

187. Deficiency of which elements makes plant more susceptible to fungal infection

a. Silicon
b. Nickel
c. Nitrogen
d. Both a and b

188. What fraction of total N in mineral soils is present in organic form

a. 25-30%
b. 40-50%
c. 60-70%
d. 95-98%

189. Technology in which plants are grown without soil is known as

a. Sand culture
b. Media Culture
c. Hydroponics
d. Biotechnology

190. The salt affected soil having low (15) ESP would be

a. Alkali soil
b. Saline soil
c. Saline-alkali soil
d. None of these

191. The exchangeable sodium ratio (ESR) can be calculated as

a. ESR = Exchangeable (Ca^{2+} + Mg^{2+})/ Exchangeable Na^{+}

b. ESR = Exchangeable Na^{+}/Exchangeable (Ca^{2+} + Mg^{2+})

c. ESR = Exchangeable Na^{+}/ Exchangeable (Ca^{2+} + K^{+})

d. ESR = Exchangeable Na^{+}/ Exchangeable (Ca^{2+} + Mg^{2+})

192. For the estimation of which one of the following is "Normal Neutral" ammonium acetate solution is used ?
   a. Available sulphur in soil
   b. Available phosphorus in soil
   c. Potassium content in plant material
   d. Available potassium content in soil

193. The term functional or metabolic nutrient was given by
   a. JV Liebig
   b. DJ Nicholas
   c. DI Arnon
   d. None

194. Which lime is referred to as burned lime or quick lime ?
   a. CaO
   b. $Ca(OH)_2$
   c. $Ca\ Mg\ (CO_3)_2$
   d. $CaCO_3$

195. Chemical formula of Gypsum is
   a. $CaSO_4$
   b. $CaSO_46H_2O$
   c. $CaSO_45H_2O$
   d. $CaSO_42H_2O$

196. Amongst oil cakes, the highest nitrogen content is in
   a. Castor cake
   b. Neem cake
   c. Coconut cake
   d. Groundnut cake

197. The Iron deficiency is found mostly in which of these soil
   a. Calcareous
   b. Acid
   c. Saline
   d. Normal

198. The process of podzolization takes place in:
   a. Warm humid climate
   b. Cold humid climate
   c. Arid climate
   d. Semi-arid climate

199. Physical condition of soil in relation to plant growth is called
   a. Tillage
   b. Till
   c. Mulch
   d. Terracing

200. 'White bud' of maize is caused due to the deficiency of

a. Zinc
b. Molybdenum
c. Nitrogen
d. Phosphorus

201. C : N ratio of normal soil is

a. 10 : 1
b. 15 : 1
c. 20 : 1
d. 30 : 1

202. pH of alkaline soil is

a. 6
b. 7
c. 8
d. 9

203. Apparent nutrient recovery (ANR) can be calculated by using of which formula

a. ANR = Nutrient removed from fertilized plot - Nutrient removed from unfertilized plot/Nutrient added x 100
b. ANR = Nutrient removed from unfertilized plot - Nutrient removed from fertilized plot/Nutrient added x 100
c. ANR = Nutrient removed from unfertilized plot - Nutrient removed from fertilized plot/ Nutrient removed from unfertilized plot x 100
d. None of these

204. Which one of the following green manuring is preferred to the crop grown in alkali soils

a. Sanai
b. Dhaincha
c. Guar
d. Cow pea

205. The concept of STCR (Targeted yield) approach in India was proposed by

a. J.S. Kanwar
b. N.P. Datta
c. N.N. Goswami
d. B. Ramamoorthy

206. Fluor apatite is a rich source of

a. Nitrogen
b. Phosphorus
c. Potassium
d. Sulfur

207. The C:N ratio of a normal mineral cultivated soil is:

a. 6 to 8
b. 10 to 12
c. 14 to 16
d. 18 to 20

208. Which one of the following is commonly used as urease inhibitor

a. Dicyandiamide
b. Thiourea
c. Urea formaladehyde
d. Phenol phosphorodiamidate

209. The ESP of sodic soil is :

a. Less than 15
b. More than 10
c. More than 5
d. More than 15

210. Rothamsted Experimental Station is situated at

a. UK
b. Germany
c. USA
d. India

211. Which soil conservation measures are recommended on steep slopes of hilly

a. Contour bunds
b. Bench Terraces
c. Graded bunds
d. Grassed waterways

212. The Critical level approach was proposed by

a. Cate and Nelson
b. J.B Bousingault
c. Theoder de Saucer
d. Arnon and Stout

213. Which of the following is not essential for plants ?

a. Calcium
b. Molybdenum
c. Iodine
d. Chlorine

214. Soils containing high organic matter normally have

a. Light colour
b. Red colour
c. Dark-brown colour
d. Yellow colour

215. Calcium sulphate ($CaSO_4.2H_2O$) is added to reclaim

a. Saline soil
b. Acid soil
c. Alkali soil
d. Calcareous soil

216. The area of milliwatershed is

a. 10-100 ha
b. 100-1000 ha
c. 1000-10000 ha
d. >10000 ha

217. A soil having available P of 30 kg/ha will be rated as

a. High
b. Medium
c. Low
d. Very low

218. Oldest surviving long term manurial trial in India is located at ...., initiated in 1909

a. Kanpur
b. Coimbatore
c. Karnal
d. New Delhi

219. Which one of the following crops is more responsive to zinc application

a. Rape seed
b. Paddy
c. Moong
d. Wheat

220. Alkali soils are generally found in

a. Temperate climate
b. Arid and semi-and climate
c. Sub-humid climate
d. Humid climate

221. On the basis of soil testing, the critical limit for 'Sulphur' in soil is

a. 05 PPM
b. 10 PPM
c. 15 PPM
d. 20 PPM

222. Percentage of nitrogen in diammonium phosphate (DAP) is :

a. 21%
b. 18%
c. 46%
d. 60%

223. The Fe-chelate suitable for application to crops grown on calcareous soils is

a. Fe-EDTA
b. Fe-EDDHA
c. Fe -DTPA
d. Fe- HEDTA

224. Which fertilizer is called 'Sona'

a. Ammonium nitrate
b. Urea
c. Diammonium phosphate
d. Calcium ammonium nitrate

225. Apatite is a group of minerals which mainly contains

a. Sulphur
b. Phosphate
c. Nitrogen
d. Manganese

226. The process that transforms the productive land into unproductive land is called

a. Desertization
b. Desertification
c. Both (a) and (b)
d. None of these

227. Leaching requirement can be estimated by formula

a. $LR = EC_{iw} / EC_{dw}$
b. $LR = EC_{dw} / EC_{iw} \times 100$
c. $LR = EC_{dw} \times EC_{iw}$
d. None of the above

228. The elements which helps in biological nitrogen fixation is

a. Mg (Magnesium)
b. Mn (Manganese)
c. Zn (Zinc)
d. Mo (Molybdenum)

229. Blossom end rot in tomato is caused by

a. Deficiency of sulpher
b. Deficiency of calcium
c. Fungus
d. None of these

230. Who is considered the father of field plot technique

a. JB Boussingault
b. Pushkarnath
c. Ramanunjan
d. Liebig

231. Browning in cauliflower occurs due to deficiency of

a. Nitrogen
b. Boron
c. Sulpher
d. Molybdenum

232. Most salt tolerant grass is

a. Karnal grass (Diplacline fusca)
b. Rhodes grass (Chloris gayann)
c. Para grass (Brachiaria mutica)
d. Bermuda grass (Cynodon dactylon)

233. .... refers to a condition in which a crop needs more of a given nutrient yet has shown no deficiency symptom

a. Bare economic optimum
b. Luxury consumption
c. Hidden hunger
d. Toxicity of nutrient

234. Which of the following is known as physiologically dry soil ?

a. Saline soil
b. Alkaline soil
c. Sandy soil
d. Calcareous soil

235. Which country have monopoly in P deposits

a. Morocco
b. Sudan
c. Russia
d. America

236. The plant species which is highly sensitive to chlorine toxicity

a. Mustard
b. Citrus
c. Tobacco
d. Onion

237. Which one of the following is not a nitrification inhibitor

a. IBDU
b. DCD
c. PMA
d. C_2H_2

238. The element present in the porphyrin structure of chlorophyll molecules are

a. Magnesium
b. Iron
c. Nitrogen
d. Magnesium and nitrogen

239. 'Annals of Agriculture' a book was written by

a. Sir Humphry Davy
b. Arthur Young
c. JS Kanwar
d. Robert Warrington

240. Which one of the following is commonly used a urease inhibitor

a. Phenyl phosphorodiamidate
b. Urea-formaldehyde
c. Nitrapyrin
d. N-serve

241. Residual sodium carbonate is calculated as

a. $(CO_3^{2-} + HCO_3^-) - (Ca^{++} + Mg^{++})$
b. $(Ca^{++} + Mg^{++}) - (CO_2^{3-} + HCO_3^-)$
c. $(Ca^{++} + Mg^{++}) - (CO_2^{3-} - HCO_3^-)$
d. $(CO_2^{3-} - HCO_3^-) - (Ca^{++} + Mg^{++})$

242. Diagnosis and recommendation integrated system (DRIS) is applicable for

a. Long duration crops
b. Short duration crops
c. Oilseed crops
d. Vegetables

243. The first N fertilizer manufactured in India was

a. Ammonium chloride
b. Ammonium sulphate
c. Urea
d. CAN

244. Biosuper is a fertilizer containing

a. Rock phosphate + elemental sulphur + S oxidizing bacteria
b. Phosphoric acid + elemental sulphur + S oxidizing bacteria
c. Ammonium sulphate + elemental sulphur + S oxidizing bacteria
d. None of these

245. Which one of the following contains maximum N

a. DAP
b. Suphala
c. Ammonium sulphate
d. Anhydrous ammonia

246. Which of the following element helps to prevent lodging in plants?

a. Nitrogen
b. Phosphorus
c. Sulphur
d. Potassium

247. Which micronutrient is most strongly bound to organic matter

a. Fe
b. Cu
c. Zn
d. B

248. Which of the following is most popular zinc fertilizer

a. Zinc sulphate
b. Zinc carbonate
c. Zinc chloride
d. Zinc EDTA

249. Factor for converting organic carbon into organic matter is

a. 1.724
b. 1.921
c. 2.724
d. 2.921

250. Who suggested that the principal nourishment of plant was water

a. Pietro de Crescenzi
b. J B Helmont
c. Francis Bacon
d. Jethro Tull

251. Which chemicals are not used as amendments for reclaiming the alkali soils ?

a. Gypsum
b. Pyrites
c. Sulphur
d. Limestone

252. The coordinated project on long term fertilizer experiment was sponsored by ICAR in

a. 1970-71
b. 1980-81
c. 1965-66
d. 1977-78

253. The DRIS approach was given by

a. Bray
b. Beaufils
c. Jackson
d. Nemeth

254. Who gave the statement 'water was the sole nutrient of plants

a. Robert Boyle
b. John Woodward
c. JR Glauber
d. JB Helmont

255. Fertilizer control order issued in the year

a. 1985
b. 1957
c. 1966
d. 1972

256. The moist productive soil of India is :

a. Alluvial soil
b. Laterite soil
c. Black soil
d. Saline soil

257. Conventional deficiency thresholds (mg/kg soil) of Zn and Fe for Indian soils are

a. 0.2 and 2.0
b. 0.6 and 4.5
c. 1.0 and 2.0
d. 3.0 and 6.0

258. Red soils are mainly confined to

a. Uttar Pradesh
b. Rajasthan
c. Tamil Nadu
d. Bihar

259. Crop logging developed by CF Clement is carried out for sugarcane in

a. Brazil
b. Hawai
c. New Zealand
d. Cuba

260. The Primary objectives of tillage is

a. Seedbed preparation
b. Provision of a good medium for plant root
c. Water infiltration and retention
d. Erosion and weed control

261. Which of the following is not a slow-release fertilizer

a. N-serve
b. LCU
c. SCU
d. NCCU

262. When organic matter is > 1.5% in soil, rated as

a. Low
b. Medium
c. High
d. Very high

263. The P content of most mineral soils is between

a. 0.02 and 0.5%
b. 0.02 and 0.5 ppm
c. 1 - 2%
d. > 2%

264. Leaching of soluble salts from either by irrigation water or rainwater is called as:

a. Alkalization
b. Salinization
c. Desalinization
d. Solodization

265. Choose the correct statement in given option

a. Zinc is a constituent of chlorophyll
b. Potassium increases disease resistance in plants
c. Copper deficiency hinders cell division in plants
d. Chlorosis between the veins is a major symptom of boron deficiency

266. Deficiency of Mo and B is more common in

a. Alkaline soil
b. Sadie soil
c. Acidic soil
d. Neutral soil

267. Who established the essentiality of Fe for plant

a. Gilbert (1880)
b. Glauber (1902)
c. Gris (1844)
d. Grigg (1905)

268. In crop logging, during growing season, plant tissue is sampled every ...... days and analysed for N, sugar, moisture and weight of the young sheath tissue

a. 15
b. 25
c. 35
d. 45

269. What kind of chemical problem exists in a soil that has white crusts where water evaporates

a. Salinity
b. Sodicity
c. Low pH
d. High pH

270. Which nutrient is a constituent of chlorophyll and chromosomes ?

a. Mg
b. Fe
c. P
d. Ca

271. Choose the correct statement in given option

a. Sulphur deficiency symptoms resemble those of phosphorus deficiency
b. Growing tips often are damaged by zinc deficiency
c. Magnesium is generally considered to be an immobile element in plants
d. Boron is not readily translocated from older leaves to meristematic tissues

272. Critical soil test level approach was given by

a. Cate and Nelson
b. Arnon and Stout
c. Larsen
d. Bray

273. Major deposits of gypsum are found in

a. Rajasthan
b. Haryana
c. Gujarat
d. Punjab

274. Which of the following cannot form chelated compounds

a. Fe
b. Mg
c. Zn
d. Mo

275. Which of the following can a plant accumulate without exhibiting any toxicity symptoms

a. B
b. K
c. P
d. N

276. A soil problem capable of being corrected simply by the addition of good quality of irrigation water is

a. Salinity
b. Aluminium toxicity
c. Low base saturation
d. Acidity

277. .... is the ratio of the increase in osmotic pressure produced by the fertilizer to that produced by the same weight of sodium nitrate ($NaNO_3$).

a. Osmotic pressure
b. Nutrient index
c. Sodium index
d. Salt index

278. Which of the following soils would be best for general plant growth if no fertilizer or lime additions are made

a. CEC=5.4; pH=5.4
b. CEC=15; pH=8.7
c. CEC=17; pH=4.1
d. CEC=22; pH=6.7

279. Boron content in irrigation water is unsafe if it exceeds

a. 1 ppm
b. 2-3 ppm
c. 2 ppm
d. >3 ppm

280. Choose the correct statement in given option

a. When fresh plant residues are added to soil they contain more nitrogen as compare to carbon
b. Soil organic material contains maximum percentage of lignin and protein
c. In organic matter dispersion, effect of calcium is more than sodium
d. Under alkali soil situation the problem is due to salt accumulation

281. In agricultural soils, percentage of the total nitrogen is found as fixed ammonium ion

a. 5%
b. 10%
c. 20%
d. 40%

282. From top 25% of effective root zone depth, how much quantity of total water extraction take place by roots

a. 10 %
b. 20 %
c. 30 %
d. 40 %

283. Law of restitution was given by

a. Olsen
b. Das
c. Liebig
d. Bray

284. Nutrient index 1.5 to 2.5 is categorized as

a. Low
b. Medium
c. High
d. None of these

285. Which of the following is not an essential plant nutrient

a. Hydrogen
b. Carbon
c. Potassium
d. Aluminium

286. What percent of potassium sulphate (K_2SO_4) is potassium [K=39 g/mol, S=32g/mol, O=16g/mol]

a. 21%
b. 25%
c. 33%
d. 45%

287. The organic soils commonly contain more than 50% organic matter by volume and at least by weight basis

a. 20%
b. 30%
c. 40 %
d. 58 %

288. Which crop is more tolerant to salinity

a. Sunflower
b. Barley
c. Cowpea
d. Groundnut

289. Which is the main source of Cadmium pollution

a. Rock phosphate
b. Muriate of potash
c. Dolomities
d. Urea

290. Plant nutrients which show deficiency symptoms on older leaves, the correct sequence is :

a. N, P, Mg, Zn
b. Ca, Mn
c. Ca, S, Cu
d. Boron, Iron

291. Increase in soil water content and soil temperature, increases the availability of which nutrient most

a. Nitrogen.
b. Phosphorus
c. Potassium
d. Boron

292. Available phosphorus of more than 25 kg/ha of soil is considered as :

a. Low
b. Medium
c. High
d. Trace

293. Which nitrogenous fertilizer industrially made first

a. $CaCN_2$
b. DAP
c. KNO_3
d. SSP

294. Which of the following forms of phosphorus is more available to plants

a. H_3PO_4
b. $H_2PO_4^-$
c. PO_4^3
d. None of these

295. Which of the following is not an essential nutrient for plant growth

a. B
b. Cl
c. Al
d. Mo

296. The permissible maximum biuret content of urea in India is

a. 0.5%
b. 1.0%
c. 1.5%
d. 2.0%

297. Zinc sulfate monohydrate contains Zn %

a. 19
b. 54
c. 21
d. 33

298. Boron content in borax (suhaga) is

a. 11 %
b. 25 %
c. 19 %
d. 39 %

299. Which of the following micronutrients are most widely deficient in Indian soils?

a. Zinc and boron
b. Zinc and iron
c. Zinc and manganese
d. Zinc and copper

300. White irregular spots between veins of upper leaves of maize plant shows the deficiency of

a. Copper
b. Molybdenum
c. Manganese
d. Boron

301. Which of the following is not a feature of watershed?

a. Hydrological unit
b. Biophysical unit
c. Socio-economic unit
d. Production unit

302. Which of the following is the important determinant of the ligand exchange reaction?

a. CEC
b. AEC
c. pH
d. Ion size

303. Cation exchange capacity [cmol(p+)/kg] of laterite and lateritic soils in India is

a. <16
b. >20
c. 20-40
d. 40-45

304. Borax is chemically known as

a. $Na_2B_4O_7\ 10H_2O$
b. $Na_2B_4O_7\ 4H_2O$
c. $Ca_2B_6O_{11}\ 5H_2O$
d. $NaCa_2B_5O_9\ 8H_2O$

305. Most of the boron is absorbed by plants in the form of

a. B
b. H_3BO_3
c. $H_2BO_3^-$
d. BO_3^{3-}

306. Minimum percentage of water soluble phosphate (as P_2O_5) in DAP is

a. 22.5
b. 32.5
c. 42.5
d. 52.5

307. 'Peat' soil are generally characterized as

a. Black, heavy and acidic
b. Red, light and acidic soil
c. Black, light and alkaline soil
d. Black, heavy and neutral soil

308. For tobacco crop, the preferred potassic fertilizer is

a. Potassium nitrate
b. Potassium chloride
c. Potassium sulphate
d. Potassium iodate

309. P content in DAP is

a. 20%
b. 36 %
c. 46%
d. 50 %

310. The boron concentration (mg/kg) unsafe for plant growth

a. 0.75 -1.5 mg/kg
b. 0.7- 0.25 mg/kg
c. 1.5 mg/kg
d. $<$ 0.2 mg/kg

311. In Mitscherlich technique plants are grown to maturity

a. Barley
b. Oat
c. Wheat
d. Sunflower

312. Diagnosis and Recommendation Integrated System (DRIS) approach was provided by

a. Mitscherlich's
b. Beaufils
c. Liebig's
d. Wallace

313.The Zn content in $ZnSO_4 7H_2O$ (Zinc sulphate)

a. 21%
b. 23%
c. 20%
d. 16%

314. On the functional basis Fe, Zn, Mn and Cu have been classified as

a. Main structural elements
b. Accessory structural elements
c. Regulators and carriers
d. Catalyst and activators

315. Which of the following oil cakes, in addition to its manurial value, also acts as a nitrification inhibitor

a. Mustard
b. Groundnut
c. Neem
d. Sesame

316. Which fertilizer act as a nitrification inhibitor and as a slow release N source ?

a. CDU
b. IBDU
c. N- serve
d. DCD

317. Deficiency of N in plant causes yellowing of

a. Upper leaf
b. Lower leaf
c. Middle leaf
d. Whole plant

318. Which of the following crops is most tolerant to salty soils

a. Cotton
b. Corn
c. Rice
d. Carrot

319. Due to deficiency of Zn causes Khaira disease in rice field, the disease caused of rice field appear

a. Necrotic
b. Yellow coloured
c. Copper brownish coloured
d. Reddish coloured

320. Calcium cynamide contains nitrogen (%)

a. 16
b. 21
c. 26
d. 33

321. Which nutrient stimulate the root growth of plant

a. N
b. P
c. K
d. Mn

322. Biuret in urea applied as a foliar spray in citrus fruits should not be more than

a. 0.25%
b. 0.50%
c. 0.75%
d. 1.50%

323. Which of the following nutrients is lost in the greatest amount by allowing exposure of manure pile to rain

a. N
b. P
c. K
d. Ca

324. Which of the plant nutrient is constituent of cell wall and essential to stability to cell membrane

a. N
b. K
c. Zn
d. Ca

325. Which of the following nutrient is constituent of chlorophyll and chromosome

a. N
b. Mg
c. Ca
d. K

326. Bone is rich source of

a. P only
b. P and Ca
c. N and K
d. P and Mg

327. The soils having medium fertility will contains organic carbon, phosphorus and potash (Available -0.C%: P_2O_5: K_2O kg/ha)

a. 0.5-1.0, 10-30, >25
b. 0.5-1.0, 22-56, >80
c. 0.5-0.75, 22-56, >56
d. 0.5-0.75, 22-56, 120-280

328. Which fertilizer has risk of fire or even explosion ?

a. Aqua NH_3
b. Ammonium nitrate
c. Potassium nitrate
d. Ammonium chloride

329. Fertility gradient approach was given by

a. BV Mehta
b. B. Ramamoorthy
c. AB Gliosh
d. Murthy

330. If requirement of 10 ton gypsum is needed to amend the soil, then element sulphur will be required

a. 0.8 T
b. 2.8 T
c. 1 .8 T
d. 2.0 T

331. The value of Von Bemmelen factor is

a. 2.430   b. 1.247

c. 2.296   d. 1.724

332. The N concentration in N sufficient plant varies

a. 1 - 5%   b. 0.5 - 1%

c. 5 -7%   d. 0.5%

333. The nutrient element regulate the opening and closing of stomata and transformation of nutrients

a. N   b. Ca

c. P   d. K

334. The safe limit of Residual Sodium Carbonate (R.S.C.) mel for the plant growth is

a. <1.25 mel   b. <2.5mel

c. <4.0mel   d. >4.0 mel

335. Which plant nutrient move from soil to the roots by root interaction, mass flow and diffusion in equal proportion ?

a. Zinc   b. Boron

c. Sulfur   d. Calcium

336. Sulphur containing amino acids in mustard are

a. Auxines

b. Cysteine, Cystine and Phytines

c. Cysteine, Cystine and Methionine

d. Cysteine, Methionine and Auxines

337. If 12 k cal energy is accumulated in two pyrophosphate bond, then which molecule stores that energy

a. ADP   b. ATP

c. $C_2H_{12}O_6$ d. Ketones

338. Heavy metals are accumulated in higher (large amount) amount when irrigated with untreated sewage

a. Wheat seeds
b. Barley seeds
c. Mustard seeds
d. Leafy vegetables

339. Which formula is used for calculate gypsum requirement (meq/100 gm)

a. GR = ESP (final) - ESP (initial)/100 x CEC
b. GR = ESP (initial)- ESP (final) /CEC x 100
c. GR = ESP (initial)-ESP (final) /100 x CEC
d. GR = ESP (initial)- ESP (final) x 100

340. Chlorosis in upland rice is due to the deficiency of

a. Zinc
b. Iron
c. Phosphorus
d. Copper

341. The ''E" value concept is normally applied to isotopic work related to

a. Zinc
b. Carbon
c. Nitrogen
d. Phosphorus

342. Primary source of phosphorus in soils with alkaline pH is

a. Gibbsitec
b. Geothite
c. Apatite
d. Gypsum

343. Stem melanosis disease occurs in wheat due deficiency of

a. Mn
b. Zn
c. Cu
d. S

344. The efficiency of rock phosphate will be more in

a. Alkaline soil
b. Neutral soil
c. Acid soil
d. Saline alkali soil

345. Plants absorb nitrogen in soil forms of

a. Ammonical
b. Nitritie
c. Nitrate
d. Ammonical and Nitrate

346. Usually the C:N ratio of fresh plant material is

a. <10:1
b. <20:1
c. <30:1
d. >30:1

347. Usually in phosphate transfer from ATP, the nutrient ion required is

a. Calcium
b. Magnesium
c. Sulfate
d. Ammonium

348. Silicon is absorbed by plants in the form of

a. Disilicic acid
b. Silicon dioxide
c. Monosilicic acid
d. Silicon

349. An element essential for microorganisms for fixing N_2 in legumes and algae is

a. Molybdenum
b. Cobalt
c. Nickel
d. Magnesium

350. Fertilizers containing Cl^- ions are not suitable for

a. Sugarcane
b. Potato
c. Wheat
d. Pigeonpea

351. The process in which NH_4^+ is converted to NO_3^- in soil is called

a. Decomposition
b. Nitrification
c. Ammonification
d. Amminisation

352. Conversion of organic N in to inorganic N is called

a. Nitrification
b. Mineralisation
c. Ammonification
d. Immobilisation

353. Downey mildew of pearlmillet may be suppressed by increased soil levels of

a. Ca
b. Cu
c. Cl
d. Mo

354. .... is a type of customised fertilizer

a. Boronated urea
b. DAP
c. Sufala 26:26:0
d. CAN

355. A fertile soil contains

a. High total amount of plant nutrients

b. High amount of available form of plant micronutrients

c. Adequate total amount of plant nutrients

d. Adequate amount of available form of plant nutrients

356. In plants, deficiency of Zn are usually associated with concentrations of less than ....

a. 6 ppm
b. 20 ppm
c. 50 pm
d. 100 pm

357. Fixation of phosphorus in the soil can be kept at minimum by maintaining pH, in the range of

a. 5-6
b. 6-7
c. 7-8
d. 8-9

358. Farmers are advised to apply urea at appropriate depth in the soil to avoid its losses through

a. Leaching
b. Adsorption
c. Volatilization
d. Fixation

359. In the soil quality index equation given by Parr *et al* [SQI = f(SP, P, E, H, ER, BD, FQ, MI)], 'H' stands for

a. Hydraulic conductivity
b. Humus content
c. Human and animal health
d. Hydryolysis

360. The 46% P_2O_5 equals to

a. 20% P
b. 30.5% P
c. 23% P
d. 26% P

361. Bronzing is associated with Fe levels greater than 300 ppm in the leaf blade of

a. Wheat
b. Rice
c. Maize
d. Sorghum

362. Ammonification of $H_4P_2O_7$ results in the formation of fertilizer called

a. Diammonium phosphate
b. Triammonium phosphate
c. Monoammonium phosphate
d. Dicalcium phosphate

363. If electrical conductivities of irrigation and drainage water are 2 and 8 dS/m, respectively, leaching requirement of the soil would be

a. 25%
b. 4%
c. 16%
d. 40%

364. Fe deficiency is mostly seen in crops growing on

a. Calcareous or alkaline soils
b. Salt affected soils
c. Acidic soils
d. Neutral soils

365. Ammonium sulphate $[(NH_4)_2SO_4]$ contain

a. 12% S
b. 18% S
c. 24% S
d. 15% S

366. What fraction of phosphorus present in inorganic fertilizers is taken up by the plants during the year of their application

a. 10-15 %
b. 20-25 %
c. 15-20 %
d. 25-30 %

367. Borax ($Na_2B_4O_7.10H_2O$) is commonly used as boron fertilizer which contains.... % boron in the water soluble form

a. 10.6
b. 5.0
c. 14.0
d. 10.0

368. The availability of iron is maximum when the soil pH ranges between

a. 7.0-8.0
b. 8.0-9.0
c. 4.0-6.0
d. 9.0-10.0

369. Among the families of crop plants, the sulphur requirement increases in the order of

a. Leguminosae > Cruciferae>Graminae
b. Graminae>Cruciferae>Leguminosae
c. Graminae>Leguminosae >Cruciferae
d. None of these

370. Auxin concentration in plants is controlled by

a. Calcium
b. Zinc
c. Magnesium
d. Boron

371. How many tonnes of compost containing 2% N needs to be applied to meet 60 kg N/ha of crop requirement. The coefficient of nitrogen mineralization is supposed to be 25%

a. 5.0
b. 7.5
c. 12.0
d. 14.0

372. The use efficiency of fertilizer N in submerged rice crop is

a. 20-30%
b. 50-60%
c. 30-40%
d. 70-80%

373. Losses through ammonia volatilization will be highest with the top dressing of

a. Ammonium chloride
b. Urea
c. Ammonium nitrate
d. Ammonium sulphate

374. The toxic substance found in fertilizer grade urea is

a. Ammonium
b. Carbonate
c. Amide
d. Biuret

375. Cadmium toxicity in the soil-plant system can be reduced to some extent by applying

a. Nickel
b. Nitrogen
c. Potassium
d. Zinc

376. What will be the leaching requirement (LR) in a soil when EC of the saturation extract is 10 dS/m at 20% reduction in crop yield The EC of irrigation water used is 1.4 dS/m

a. 6.5%
b. 7.5%
c. 7.0%
d. 8.0%

377. The purpling along the leaf edges is the characteristic deficiency symptom of

a. P b. K

c. B d. Fe

378. A wheat crop in a long term rice-wheat experiment, supplied with recommended N fertilizer, shows typical N deficiency symptoms after a heavy rainfall It could be due to

a. Lower air-filled porosity b. Fixation of N in the soil

c. Leaching of applied N fertilizer d. Higher C:N ratio

379. Molybdenum is adsorbed strongly by

a. Ca and Mg oxides b. C compounds

c. Iron and Al oxides d. $CaCO_3$

380. According to Ministry of Agriculture report (2018-19), the average per hectare consumption of NPK fertilizers was highest in the state of

a. Punjab b. Telengana

c. Puducherry d. Haryana

381. The study of soils in relation to plant growth is called as

a. Pedology b. Edaphology

c. Geology d. Physiology

382. The maximum percent of biuret in urea as specified in "The Fertilizer Control Order' is

a. 1.5 b. 2.5

c. 2.0 d. 3.5

383. The primary constituent of chlorophyll (about 20%) is

a. Nitrogen b. Phosphorus

c. Calcium d. Magnesium

384. Which element is considered as energy currency for the plant

a. N b. P

c. K d. Mo

385. The land capability classes suited to cultivation is

a. V to VIII
b. I to III
c. I to IV
d. VI to VIII

386. The nutrient concentration range in which added nutrient will not increase yield but can increase nutrient concentration is referred as

a. Deficient range.
b. Critical range
c. Luxury consumption
d. Toxic range

387. The micronutrient available under high pH is

a. Iron
b. Zinc
c. Boron
d. Molybdenum

388. Radiation most useful for photosynthesis is

a. 300-900 nm
b. 300-400 nm
c. 400-700 nm
d. 400-600 nm

389. "Increase in yield of a crop as a result of increasing a single growth factor is proportional to the decrement from the maximum yield obtainable by increasing the particular growth factor" was described by

a. Liebig
b. Bray
c. Spillman
d. EA Mitscherlich

390. Bray has modified the mitscherlich equation to

a. G = fx2, x3 xn)
b. dy/dx = (A -y) C
c. Y = A (1-C)
d. log (A-y) = log A - C1b - cx

391. The mineral source of boron in soil is

a. Apatite
b. Pyrite
c. Tourmaline
d. Bentonite

392. The N serve is a

a. Nitrification inhibitor
b. Nitrogen fixer
c. Nitrification enhancer
d. Urease inhibitor

393. The concept of DRIS was developed by

a. Cate and Nelson
b. Beaufils
c. Lindsay and Norvell
d. Page and Miller

394. Scab of Irish potatoes, pox of sweet potatoes, and black root rot of tobacco are favoured by

a. Acidic. conditions
b. Highly alkaline
c. Neutral to alkaline conditions
d. Extremely acidic condition

395. The inherent capacity of a soil to supply essential nutrients to plants in adequate amount and in right proportion for their optimum growth is called

a. Soil fertility
b. Optimum fertilizer rates
c. Efficient use of fertilizer
d. Integrated nutrient management

396. The behaviour of the plant in relation to day length is termed

a. Thermoperiodism
b. Photoperiodism
c. Heliotropic
d. None of above

397. Which of the following is commonly referred to as muriate of potash

a. Potassium nitrate
b. Potassium sulphate
c. Potassium chloride
d. Potassium silicate

398. Indian agriculture is operating at a negative balance of plant nutrients

a. 2 - 5 mt
b. 8 - 10 mt
c. 10 - 15 mt
d. 20 - 25 mt

399. In 1843, Agricultural Experiment Station at Rothamsted, England was started by

a. Liebig
b. Blackman
c. Lawes (Sr Gilbert)
d. Benjamin Franklin

400. In acid soil, which one may be occurring

a. Aluminium toxicity
b. Magnesium toxicity
c. Nitrate toxicity
d. Chloride toxicity

401. It has been observed that soil pH increases in

a. Summer
b. Winter
c. Zaid
d. Post monsoon season

402. Anaerobic environment of paddy soil is responsible for which of the following process that result in the gaseous loss of fertilizer nitrogen

a. Ammonification
b. Nitrification
c. Denitrification
d. Volatilization

403. The law of the minimum, was given by Liebig in

a. 1842
b. 1852
c. 1862
d. 1875

404. Which country has largest producers of K fertilizer

a. India
b. Canada
c. U.S.A.
d. China

405. 'Hidden hunger' means:

a. Deficiency symptoms are seen when the nutrient is deficient
b. Severe yield reduction may occur without appearance of deficiency symptoms
c. The nutrient is not deficient but apparently seems to be deficient
d. Visual deficiency symptoms are suppressed by other elements

406. Absorption of ions in plants occurring with the aid of metabolic energy is termed as

a. Passive absorption
b. Active absorption
c. Diffusive absorption
d. Mass flow absorption

407. Molasses is a by product of which industry

a. Textile industry
b. Sugar industry
c. Tanning industry
d. Paper industry

408. "Mridaparikshak", a MINILAB for soil testing is developed by

a. ICAR-IISS, Bhopal
b. ICAR-IARI, New Delhi
c. ICAR-CAZRI, Jodhpur
d. ICAR-CRIDA, Hyderabad

409. 'Biochar' is used for

a. Slow release fertilizer.
b. Bio-fertilizer for N_2-fixation
c. Carbon sequestration
d. Ameliorant for Alkali Soil

410. Which of the following is commonly referred to as muriate of potash

a. Potassium nitrate
b. Potassium chloride
c. Potassium sulphate
d. Potassium silicate

411. For applying 100 kg of nitrogen, how much urea would one use

a. 45kg
b. 111kg
c. 222kg
d. 333kg

412. The concept of nutrient mobility in soils was first proposed by

a. R.H. Bray
b. M.L Jackson
c. E. Mitscherlich
d. E.W. Russell

413. 'Horse hoeing husbandry' was written by

a. Liebig
b. Arthur Young
c. Baker
d. Jethro Tull

414. In India, which fertilizer manufactured first

a. Urea
b. DAP
c. SSP
d. ASN

415. Which one is a major component of organic P in soils

a. Nucleic acids
b. Phospholipids
c. Inositol phosphate
d. ATP

416. .... suggest that saltpeter (KNO_3), and not water, was the 'principle of vegetation'

a. JR Glauber
b. Jethro Tull
c. Johnwoodward
d. John Mayow

417. Soil test based fertilizer recommendation for targeted yield of crops was first introduced in India by

a. J.S. Kanwar
b. N.P. Datta
c. B. Ramamoorthy
d. R.V. Tamhane

418. Phosphate-fixation is minimum at soil pH between

a. 8 and 9
b. 7 and 8
c. 6 and 7
d. 5 and 6

419. The phosphate-fixation capacity is in the different order soils

a. Red>medium black>hevay black>grey bronwn >alluvial
b. medium black>hevay black>grey bronwn >alluvial > Red
c. hevay black>grey bronwn >alluvial >Red>medium black
d. grey bronwn >alluvial >Red>medium black>hevay black

420. The first compound formed on urea hydrolysis is

a. Ammonium carbonate
b. Ammonium carbamate
c. Anhydrous ammonia
d. Ammonium hydroxide

421. Uniform chlorosis of the youngest growth is a symptom of

a. K deficiency
b. S deficiency
c. Mg deficiency
d. Fe deficiency

422. Which one is not a resource conservation technology (RCT)

a. Zero tillage
b. Puddling
c. Laser leveling
d. Bed planting

423.Real-time N management refers to

a. N supply as per plant demand
b. N supply as per soil test
c. N supply as per fertilizer schedule
d. N supply at different growth stages

424. Soil fertility and fertilizers' was written by

a. NC Brady
b. Tisdale et al.
c. MM Rai
d. None

425. Amount of micronutrients required by various crops can be arranged in an order:

a. Fe>Mn>Zn>Cu>Mo
b. Mn>Zn>Fe>Cu>Mo
c. Fe>Zn>Mn>Cu>Mo
d. Mo>Cu >Zn>Mn>Fe

426. Choose the correct statement in given option

a. Excess copper may enhance the utilization of molybdenum
b. Excess potassium enhances manganese uptake
c. Heavy nitrogen fertilization intensifies copper deficiency
d. Excess of zinc, manganese and copper reduces the iron deficiency

427. Choose the correct statement in given option

a. In triple super phosphate 46-48 percent phosphorus is as monocalcium phosphate or water soluble phosphate
b. In alkaline soils, iron and aluminium are present in appreciable, quantity
c. From rock phosphate, availability of phosphorus is more in alkaline reaction soils
d. Ammonium ion containing fertilizers are difficult to dissolve in water

428. Choose the correct statement in given option

a. At low temperature both growth and decomposition rate of plant is more
b. Finer the soil texture lesser will be the carbon content
c. Soils ploughed repeatedly contain less carbon content
d. Loss of soil carbon occurs in legume containing crop rotation

429. During which, pH of mineral soil declines

1. Process of organic matter decomposition
2. Drying of soil
3. In Summer
4. In Winter

Select the correct answer using the codes:

a. (1), (2) and (3) b. (1), (2)and (4)
c. (1) and (3) d. (1) and (4)

430. Leaching of salts from a saline sodic soil is accompanied by

1. Flocculation of clays
2. Swelling of the colloids
3. Reduction in soil permeability
4. Development of excessive alkalinity

Select the correct answer using the codes:

a. (1), (3) and (4) b. (2), (3) and (4)

c. (1), (2) and (3) d. (2) and (3)

431.What are the fates of nitrates in the soil

1. Leaching 2. Denitrification

3. Nitrification 4. Plant uptake

Select the correct answer using the codes:

a. (1), (2) and (3) b. (1), (2) and (4)

c. (3) and (4) d. (2) and (3)

432. Land capability class I soils are

1. Deep and well drained
2. Have moderate erosion hazards
3. Good nutrient supplying power
4. Their productivity can be managed with ordinary crop management practices

Select the correct answer using the codes:

a. (1), (3) and (4) b. (2), (3) and (4)

c. (1) and (3) d. (2) and (4)

433.Which of the following are associated with the zinc nutrition

1. Itai-Itai disease 2. White bud of maize

3. Khaira disease 4. Little leaf of cotton

Select the correct answer using the codes:

a. (1), (2) and (4) b. (2), (3) and (4)

c. (2) and (3) d. (2) and (4)

434. Which of the following are associated with the manganese nutrition

1. Grey specks of oats 2. Crinkle leaf of cotton

3. Whiptail of cauliflower 4. Speckled yellow of sugar beet

Select the correct answer using the codes:

a. (1), (2) and (4) b. (1), (2) and (4)

c. (1), (3) and (4) d. (1) and (4)

435. Loss of potassium by leaching increases due to

1. Low cation exchange capacity of sandy soils.
2. Relatively higher proportion of expanding type of clay
3. Soils with high degree of base saturation
4. Alternate wetting and drying of rice fields

Select the correct answer using the codes:

a. (1), (2) and (4) b. (1), (2) and (3)

c. (1) and (4) d. (1) and (3)

436. Consider the following statements :

1. Zinc ion is a cofactor for the enzyme carbonic anhydrase.
2. A common symptom of zinc deficiency in plants is leaf mottling.

Select the correct answer using the codes:

a. (1) only b. (2) only

c. Both (1) and (2) d. Neither (1) non (2)

437. Consider the following statements :

1. Some plants exhibit iron chlorosis when grown in calcareous soils.
2. High concentration of calcium ions in soil prevent the absorption of iron by root hairs.

Select the correct answer using the codes:

a. (1) only b. (2) only

c. Both (1) and (2) d. Neither (1) nor (2)

438. Consider the following statements :

1. Diffusion is the dominant mechanism for transporting soil Zn^{2+} to plant roots.
2. Zinc solubility in soil solution decreases 100 fold for each unit increase in pH.

Select the correct answer using the codes:

a. (1) only b. (2) only

c. Both (1) and (2) d. Neither (1) nor (2)

439. The acidic soils are deficient in which of the below nutrient elements

1. Ca 2. Zn

3. P 4. Mg

Select the correct answer using the codes:

a. 1, 2 & 3 only b. 2, 3 & 4 only

c. 1, 3 & 4 only d. 1, 2 & 4 only

440. Consider the following :

1. Phosphorus availability is one of the major constraints to plant growth in nature.
2. Phosphorus has low solubility and high sorption capacity in soil.

Select the correct answer using the codes:

a. 1 only b. 2 only

c. Both 1 and 2 d. Neither 1 nor 2

441. Match List I with List II

a. DRIS concept 1. Parkar

b. Salt index 2. Fried and Dean

c. A value 3. Beckett

d. Nutrient index 4. Puri

442. Match List I with List II

a. Integrated soil test approach 1. Larsen

b. Biogas 2. Desai & Biswas

c. L value 3. Colwell

d. Law of limiting factor 4. Lundegardh

443. Match List I with List II

a. Deflorination 1. Potassic fertilizers

b. Floatation 2. Nitrogenous fertilizers

c. Anti-caking	3. Biochar
d. LCC	4. Conditioners
e. Pyrolysis	5. Phosphatic fertilizers

444. Match List I with List II

a. Infrared thermometer.	1. Wind velocity
b. Psychrometer.	2. Stable Isotopes
c. Spectro-radiometer	3. Refiectance
d. Anemometer	4. Canopy temperature
e. Mass spectrometer	5. Relative humidity

445. Match List I with List II

a. Environmental Soil Physics	1. C.E. Marshall
b. Physical Chemistry and Mineralogy of Soil	2. D. Hillel
c. Soil Conditions and Plant Growth	3. W.L. Lindsay
d. Soil Fertility: Theory and Practice	4. E.W. Russell
e. Chemical equilibria in Soils	5. J. S. Kanwar

446. Match List I with List II

a. pH meter	1. Micronutrients
b. EC bridge	2. Phosphorus
c. Flame photometer	3. Specific conductance
d. Spectrophotometer	4. Active acidity
e. AAS	5. Na and K

447. Match List I with List II

a. Urea	1. 33%
b. Ammonium sulphate	2. 82%
c. Ammonium nitrate	3. 21%
d. Diammonium phosphate	4. 46%
e. Anhydrous ammonia	5. 18%

448. Match List I with List II

List I	List II
a. Land Coordinates	1. Mineralization and immobilization
b. Land use	2. Soil capability classification
c. C:N ratios	3. Environmental
d. Land slides and soil erosion	4. Hilly areas
e. Soil and water pollution	5. GPS

449. Match List I with List II

List I	List II
a. Poultry manure	1. Calcium
b. Pyrites	2. Biofertilizer
c. Gypsum	3. Carbon
d. Compost	4. Organic manure
e. Azolla	5. Sulphur

450. Match List I (Elements) with List II (Tissue concentration PPM)

List I	List II
a. Fe	1. 10-300
b. Mn	2. 2-100
c. Zn	3. 0.2-5
d. Cu	4. 10,000-1,00,000
e. Mo	5. 20-4,000

451. Match List I (Institutes) with List II (Places)

List I	List II
a. Spices Board	1. Cochin
b. Council for Leather Exports	2. New Delhi
c. Central Silk Board	3. Chennai
d. Central Cottage Industries Corporation	4. Bangalore

452. Match List I (Concept) with List II (Scientists)

a. Theory of "optima and limiting factors"	1. Wilcox
b. Concept of "critical percentages limiting factors"	2. Bennet and Clark
c. Inverse Yield-Nitrogen Law	3. F.T. Blackman
d. Indentified ABA	4. Macy

453. Match List I with List II

a. A Value	1. Epstein
b. Q/I relationship of P	2. Fried and Dean
c. Terminal Velocity Low	3. G G Stokes
d. Carrier Theory	4. Schofield

454. Match List I with List II

a. H_2O_2	1. Elements highly mobile in plant
b. Phosphate potential	2. Less mobile nutrients in soil
c. N, P, K	3. Removal of humus from soil sample
d. Ca, Mg, Cu	4. ½ pCa+pH_2PO_4

455. Match List I (Concept) with List II (Scientists)

a. Sodium Absorption Ratio	1. Page & Baver (1940)
b. Low of Restitution	2. Bernstein and Francois
c. Lattice Hole Theory	3. Richards
d. Lime Requirement	4. Juston von Liebig

456. Match List I (Elements) with List II (Function)

a. N	1. Electron transport system
b. P	2. Has no organic form
c. K	3. Nucleotide synthesis
d. S	4. Protein synthesis
e. Fe	5. Amino acids

ANSWERS (Soil Fertility)									
1.b	2.a	3.a	4.b	5.a	6.c	7.c	8.c	9.b	10.d
11.c	12.c	13.b	14.c	15.a	16.d	17.c	18.b	19.b	20.a
21.b	22.b	23.a	24.a	25.c	26.a	27.c	28.d	29.b	30.a
31.c	32.a	33.c	34.a	35.d	36.c	37.c	38.a	39.a	40.a
41.a	42.a	43.d	44.a	45.b	46.b	47.d	48.b	49.c	50.a
51.b	52.a	53.d	54.c	55.b	56.b	57.c	58.b	59.a	60.a
61.b	62.b	63.b	64.a	65.a	66.d	67.d	68.d	69.c	70.b
71.a	72.c	73.c	74.b	75.a	76.a	77.c	78.a	79.b	80.d
81.b	82.c	83.c	84.c	85.b	86.c	87.b	88.c	89.a	90.b
91.c	92.d	93.c	94.c	95.a	96.c	97.c	98.c	99.a	100.b
101.b	102.a	103.d	104.b	105.a	106.d	107.c	108.c	109.c	110.d
111.b	112.a	113.a	114.a	115.b	116.d	117.c	118.a	119.c	120.a
121.b	122.d	123.a	124.c	125.d	126.a	127.d	128.b	129.b	130.d
131.b	132.d	133.c	134.b	135.b	136.d	137.b	138.b	139.a	140.b
141.c	142.d	143.d	144.b	145.b	146.a	147.b	148.c	149.a	150.c
151.d	152.d	153.c	154.c	155.c	156.b	157.b	158.b	159.d	160.c
161.a	162.c	163.c	164.b	165.d	166.c	167.d	168.b	169.d	170.a
171.a	172.a	173.d	174.c	175.b	176.d	177.d	178.a	179.c	180.a
181.d	182.d	183.b	184.c	185.a	186.d	187.d	188.d	189.c	190.b
191.d	192.d	193.b	194.a	195.d	196.d	197.b	198.b	199.b	200.a
201.a	202.d	203.a	204.b	205.d	206.b	207.b	208.d	209.d	210.a
211.a	212.a	213.c	214.c	215.c	216.c	217.a	218.b	219.b	220.b
221.b	222.b	223.b	224.d	225.b	226.c	227.a	228.d	229.b	230.a
231.b	232.a	233.c	234.a	235.a	236.c	237.d	238.d	239.b	240.a
241.a	242.a	243.b	244.a	245.d	246.d	247.b	248.a	249.a	250.c
251.d	252.a	253.b	254.d	255.b	256.a	257.b	258.c	259.b	260.a
261.a	262.c	263.a	264.c	265.b	266.c	267.c	268.c	269.a	270.a
271.d	272.a	273.a	274.b	275.b	276.a	277.d	278.d	279.d	280.b
281.b	282.d	283.c	284.b	285.d	286.d	287.a	288.b	289.a	290.a
291.b	292.c	293.a	294.b	295.c	296.c	297.d	298.a	299.a	300.c
301.d	302.a	303.a	304.a	305.b	306.c	307.a	308.a	309.a	310.c
311.b	312.b	313.a	314.d	315.c	316.d	317.b	318.a	319.c	320.b
321.b	322.a	323.a	324.d	325.b	326.b	327.d	328.b	329.b	330.c
331.d	332.a	333.d	334.a	335.a	336.c	337.c	338.d	339.c	340.b
341.d	342.c	343.c	344.c	345.d	346.d	347.c	348.c	349.a	350.b
351.b	352.b	353.c	354.a	355.d	356.b	357.b	358.c	359.c	360.a
361.b	362.b	363.a	364.a	365.c	366.a	367.a	368.c	369.c	370.b
371.c	372.c	373.b	374.d	375.d	376.c	377a	378.c	379.c	380.b
381.b	382.a	383.d	384.b	385.c	386.c	387.d	388.b	389.d	390.d
391.c	392.a	393.b	394.c	395.a	396.b	397.c	398.b	399.c	400.a
401.b	402.c	403.c	404.b	405.b	406.b	407.b	408.a	409.c	410.b
411.c	412.a	413.d	414.c	415.c	416.a	417.c	418.c	419.b	420.b

ANSWERS (Soil Fertility)									
421.b	422.b	423.a	424.b	425.a	426.d	427.a	428.c	429.c	430.b
431.b	432.a	433.b	434.b	435.c	436.c	437.c	438.c	439.c	440.c
441. a.3 b.4 c.2 d.1					442. a.3 b.2 c.1 d.4				
443. a.1 b.5 c.4 d.2 e.3					444. a.4 b.5 c.2 d.1 e.3				
445. a.2 b.1 c.4 d.5 e.3					446. a.4 b.3 c.5 d.2 e.1				
447. a.4 b3. c.1 d.5 e.2					448. a.5 b.2 c.1 d.4 e.3				
449. a.4 b.5 c.1 d.3 e.2					450. a.4 b.5 c. 1 d.2 e.3				
451. a.1 b.3 c.4 d.2					452. a.3 b.4 c.1 d.2				
453. a.2 b.4 c.3 d.1					454. a.3 b.4 c.1 d.2				
455. a.3 b.4 c.1 d.2					456. a.4 b.3 c.2 d.5 e.1				

3
Soil Microbiology

1. Amoeba are unicellular organisms that feed on

    a. Bacteria and algae
    b. Bacteria and fungi
    c. Fungi and algae
    d. Nematode and fungi

2. Vermi-compost is made by use of

    a. Earthworms
    b. House fly
    c. Termites
    d. Birds

3. Vascular Arbuscular Mycorrhiza (VAM) increase uptake of

    a. Fe
    b. Cu
    c. Zn
    d. Ni

4. Azolla is a

    a. Algae
    b. Fern
    c. Tree
    d. Fruit

5. According to Bureau of Indian standard, what is the cell content of rizobium and azotobacter in biofertilizer?

    a. 10^8/g and 10^7/g
    b. 10^7/g and 10^8/g
    c. 10^5/g and 10^6/g
    d. 10^9/g and 10^6/g

6. Father of soil microbiology is

    a. S. N. Winogradsky
    b. A. Wakaman
    c. Liebing
    d. Beijerinck

7. Which of the following are included in soil microflora

    a. Bacteria, fungi, actinomycetes and algae
    b. Only bacteria and fungi
    c. Only fungi and actinomycetes
    d. Bacteria, fungi, actinomycetes and nematodes

8. Biological Nitrogen Fixation (B.N.F.) is catalized by

a. Cu	b. Zn
c. B	d. Mo

9. Which group of microorganisms have highest biomass in soil

a. Bacteria	b. Actinomycetes
c. Fungi	d. Algae

10. In nitrification which organism convert Nitrite to Nitrate is

a. Azotobacter	b. Nitrobacter
c. Nitrosomonas	d. Nitrogenase

11. Which one is the aerobic non-symbiotic nitrogen fixing bacteria?

a. Azotobacter	b. Clostridium
c. Rhizobium	d. Beijerinckia

12. Who discovered the atmosphere nitrogen fixation by plants?

a. Hellriegel	b. M.W. Beijerinck
c. Arnon	d. Nicholas

13. Which bacteria involves for the reduction of sulfur in soil

a. Thiobacillus	b. Desulphovibrio
c. Pseudomonas	d. Bacillus

14. Micro organism grow at temperature below 5°C are called

a. Thermophiles	b. Mesophiles
c. Psychrophiles	d. None of these

15. Which of the following is a greenhouse gas

a. Ammonia	b. Chlorine
c. Oxygen	d. Methane

16. Which of the following microorganism grows well in logged soils

a. Clostridium	b. Azotobacter
c. Rhizobium	d. Radiobacter

17. Microbial conversion of nitrate into chemical nitrogen is called
    a. Denitrification
    b. Nitrate reduction
    c. Nitrogen fixation
    d. None of these
18. Nitrification is a process of
    a. Reduction
    b. Hydrolysis
    c. Oxidation
    d. None of these
19. Fraction of organic matter comparatively resistant to microbial degradation
    a. Protein
    b. Lignin
    c. Cellulose
    d. Hemicelluloses
20. Which of the fallowing is an associative N-fixer
    a. Rhizobium
    b. Azotobacter
    c. Azospirillum
    d. Beijerinckia
21. Nitrification in the soil takes place by
    a. Oxidation process
    b. Biological oxidation
    c. Reduction process
    d. Both oxidation and reduction process
22. Dominant organisms found in a compost pit are
    a. Thermophilic
    b. Mesophilic
    c. Psychrophilic
    d. Psychrophilic & Mesophilic
23. Atmosphere nitrogen fixations by legume crop with rizobium discovered by
    a. M.W. Beijerinck
    b. Boussiongault
    c. Arnon
    d. Nicholas
24. The blue-green algae belong to
    a. Cyanophyta
    b. Xanthophyta
    c. Chlorophyta
    d. Basillariophyta
25. The ability of BGA to fix N_2 was first demonstrated by
    a. P.K. De
    b. R. N. Singh
    c. J.S. Kanwar
    d. N.P. Datta

26. Rust mottles are common in
    a. Well drained soil
    b. Moderately drained soil
    c. Poorly drained soil
    d. Normal soil
27. A peculiar structure *heterocyst* is found in
    a. Rhizobium sp.
    b. Actinomycetes
    c. Cyanobacteria
    d. Fungi
28. Rhizosphere is dominated by
    a. Gram positive spore forming bacteria
    b. Gram negative spore forming bacteria
    c. Gram negative non-spore forming bacteria
    d. Gram positive non-spore forming bacteria
29. The toxicity of oxygen for strict anaerobe is due to the production of
    a. Hydroxyl radical
    b. Superoxide radical
    c. Hydrogen peroxide
    d. Cyanide
30. The specificity between legume host and Rhizobium spp. is governed by
    a. Flavonoids
    b. Tryptophan
    c. Polysaccharides
    d. Indole acetic acid
31. The microbes which can survive at the temperature 25-30°C is called
    a. Mesophylic
    b. Thermoduric
    c. Thermopillic
    d. Psychrophylic
32. Some microorganisms produce organic ligands to chelate iron in the soil and make it available to the plants. These organic ligands are known as
    a. Porphyrins
    b. Siderophores
    c. Chromophores
    d. Ferrodoxins
33. Biofertilisers are
    a. Are the products containing living cells of different types of micro-organisms
    b. Mixture of vermicomposts and inorganic fertilizers
    c. Mixture of vermicomposts and organic manures
    d. Mixture of organic manures end inorganic fertilizers

34. During decomposition of plant materials, the protein are hydrolysed in to

 a. Polypeptides b. Humin

 c. Quinines d. Humic acid

35. The nomenclature of Actinomycetes derives from a greek word Actinos means

 a. Net b. Ray

 c. Thread d. Hyphae

36. Which is not genera of blue-green algae

 a. Anabaena b. Paeudomona

 c. Nostoc d. Calothrix

37. Mineralization of organic sulphur in to sulphate is strictly a

 a. Biological process b. Thermodynamic process

 c. Physiological process d. Immobilization process

38. VAM is

 a. A microbial inoculant occur in plant roots

 b. A myeorrhiza occur in plant

 c. A phosphate solubilising micro-organism

 d. A myecorrhiza occurs as mutuality symbiosis between plant roots and soil fungi

39. Nitrogenase is an enzyme associated with

 a. Nitrogen oxidation b. Nitrate reduction

 c. Nitrogen fixation d. Nitrogen absorption

40. Exclusive source of carbon for autotrophic soil population is

 a. Glucose b. Methane

 c. Carbon dioxide d. Carbonates

41. Mycorrhizae is a symbiotic association between

 a. Bacteria and plant roots b. Fungi and bacteria

 c. Fungi and roots of higher plants d. Two species of fungi

42. The microorganism that is considered most important as an indicator of soil quality is

a. Bacteria b. Fungi

c. Actinomycetes d. Algae

43. Nitrogen reduction in symbiotic system takes place at

a. Bacteroids b. Membrane envelops

c. Nodule surface d. Root surface

44. Which of the following crop has determinate type of root nodules

a. Chickpea b. Pea

c. Soybean d. Pigeonpea

45. Methanogenesis in water-logged soil takes place at Eh

a. < -200 mV b. > - 200 mV

c. > + 200 mV d. None of them

46. Which of the following element plays a role in biological nitrogen fixation ?

a. Copper b. Zinc

c. Boron d. Molybdenum

47. Net immobilization of nitrogen occurs when C:N ratio of organic matter is

a. <20:1 b. 20-30:1

c. >30:1 d. >300:1

48. Around to ... soil of active root zone of the growing plants is designated as rhizosphere

a. 2 cm b. 2 mm

c. 4 cm d. 2 inch

49. Humic acid isolated from soils is

a. Aromatic with high molecular weight

b. Aromatic with low molecular weight

c. Aliphatic with high molecular weight

d. Aliphatic with low molecular weight

50. Which amino acid does not contain sulphur

a. Cystine
b. Cystein
c. Methionine
d. Arginine

51. Aerobic conditions can be maintained in a compost pile by

a. Frequent turning
b. Compacting the pile
c. Overwatering
d. Both (a) and (c)

52. Out of the following, which cropping system helps in better carbon sequestration

a. Rice-wheat
b. Silvi-pastoral
c. Agro-horticulture
d. Agro-forestry

53. Azolla can fix nitrogen in soil

a. 10-20 kg N/ha/year
b. 30-40 kg N/ha/year
c. 50-60 kg N/ha/year
d. 80-100 kg N/ha/year

54. Which one of the following microorganisms is responsible for maximum nutrient cycling in the soil

a. Fungi
b. Algae
c. Bacteria
d. Actinomycetes

55. Stem nodulating Rhizobium spp. is

a. Sino-rhizobium
b. Bradyrhizobium
c. Azo-rhizobium
d. None of these

56. Which of the following cropping system helps in producing more of methane as GHGs

a. Direct-dry seeded rice-wheat
b. Direct-wet seeded rice-wheat
c. Maize-wheat
d. Soybean-wheat.

57. Humus forming fungi includes which genera

a. Trichodenna
b. Alternaria
c. Aspeigillus
d. Both (b) and (c)

58. Examples of green house gases include

a. Methane
b. Nitrous oxide
c. Oxygen
d. Both (a) and (b)

59. Oxidation of sulphur to sulphate is carried out by

a. Bacillus spp. b. Azotobacter spp.

c. Asperzillus spp. d. Thiobacillus spp

60. Which of the following has the lowest C/N ratio

a. Alfalfa b. Humus

c. Manure d. Bacteria

61. The plant disease caused by an algae is

a. Blue mold of tobacco b. Red rust of tea

c. Cell wall d. Aster yellows

62. Prokaryotes in soil include the following groups of organisms

a. Fungi, Actinomycetes, Bacteria, Algae

b. Bacteria, Archaea, Actinomycetes, BGA

c. Protozoa, Bacteria, Fungi, Archaea

d. Algae, Protozoa, Fungi, Bacteria

63. Aerobic bacteria can assimilate only.......% of the carbon during organic matter decomposition

a. <1.0% b. 5-10%

c. 20-40% d. >40%

64. Which organic acid produced by microbes is more effective in solubalization of insoluble inorganic phosphates

a. Propionic acid b. Lactic acid

c. Tartaric acid d. Keto-glutaric acid

65. If 0.1 ml suspension of 10^4 dilution of a fresh soil having 10% moisture formed 45 fungi colonies, the fungal population in soil would he

a. 4.5×10^6 cfu/g b. 4.5×10^5 cfu/g

c. 5.0×10^6 cfu/g d. 5.0×10^5 cfu/g

66. One unit change in pH at 25°C is equivalent to a change in potential difference of

a. 59.10 V b. 5.91 V

c. 0.591 V d. 0.059 V

67. By which method soil texture particles is perfectly determined

a. Field method b. Use of textural triangle

c. Hydrometer method d. International pipette method

68. If the average cell composition of microflora is taken as 50% C and 5% N, calculate the amount of C going to be assimilated front a substrate containing 50 kg N

a. 5 kg b. 50 kg

c. 500 kg d. 250 kg

69. Lignin is a polymer of

a. Sugar alcohol b. Polyphenols

c. Aromatic alcohol d. Galactosamines

70. Nitrobacter is

a. Heterotroph b. Nitrite former

c. Autotroph d. Symbiotic N fixer

71. Which one of the following organisms is an obligate chemoautotroph

a. Frankia b. Nitrosomonas

c. Rhizobium d. Anabaena

72. In nitrification, the oxidation state of N changes from

a. -3 to +3 b. +3 to -5

c. -3 to +5 d. +3 to -3

73. Suppose that 100 kg of green manure containing 40% C and 2% N is decomposed by bacteria with C to N ratio of 5:1 and C assimilation efficacy of 5%. What will be the quantity of N mineralized or immobilized from the addition of green manure

a. 0.4 kg N immobilized b. 1.6 kg N mineralized

c. 2 kg N immobilized d. 2 kg N mineralized

74. 'BIOLOG' is used for determining

a. Genetic finger prints of soil microorganisms

b. Taxonomic changes of soil microorganisms

c. Metabolic finger prints of soil bacteria

d. All of the above

75. The ratio of microbial biomass C to organic carbon in soils is termed as

a. Microbial quotient b. Metabolic quotient

c. Mineralization index d. Labile C

76. Soil that contains 10^3 bacterial cells per gram is treated with organic matter. Bacteria in the soil can double every 4 hours. The number of bacterial cells after 36 hours will be

a. $4x10^6$ b. $5.12x10^5$

c. $8.2x10^5$ d. $9x10^3$

77. Main factor which influence the decomposition of organic matter is:

a. pH b. C/N ratio

c. Temperature d. Biodegradability of compound

78. Which of the following shows edotrophic mycorrhiza

a. Boletus b. Glomus

c. Amanita d. Rhizoctonia

79. In green houses which of the following gas not found abundantly

a. CO_2 b. CH_4

c. H_2O d. O_3

80. The most common pathway for assimilation of fixed N_2 in legumes is

a. GS-GOGAT b. GDH

c. ED d. EMP

81. Which of the following is the main cause for Eutrophication

a. Nitrogen b. Sulphur

c. Phosphorus d. Micronutrients

82. The major source of negative charge on humus colloids is due to

a. Carboxylic (-COOH) groups b. Enolic (-OH) groups

c. Phenolic groups d. All of these

83. Humus fraction lightest in colour is
    a. Fulvic acid
    b. Humic acid
    c. Humin acid
    d. None of these
84. Azolla is in symbiotic association with
    a. Bryphytes and blue green algae
    b. Fern and blue green algae
    c. Nostoc and blue green algae
    d. Fungi and blue green algae
85. Which organism fixes atmospheric nitrogen
    a. Azotobacter
    b. Nitrobacter
    c. Nitrosomonas
    d. Nitrosococcus
86. What is formed in aminization
    a. NO_3^-
    b. NO_2^-
    c. NH_4^+
    d. $R\text{-}NH_2$
87. Actinomycetes belong to
    a. The fungi
    b. Eukaryote
    c. Mycelia sterilia
    d. None of these
88. Which one of the following microorganisms is responsible for maximum nutrient cycling in the soil
    a. Fungi
    b. Algae
    c. Bacteria
    d. Actinomycetes
89. Easily oxidizable organic matter sometimes degrades native soil organic carbon. This is due to
    a. Synergistic effect
    b. Mutualistic effect
    c. Priming effect
    d. Sequestration effect
90. Mycorrhiza is associated with :
    a. Stem
    b. Leaf
    c. Root
    d. Flower
91. The process of biological conversion of ammonia into nitrate is known as :
    a. Ammonification
    b. Nitrification
    c. Humification
    d. Denitrification

92. Ammonia is transformed to nitrate in the soil by
    a. Fungi
    b. Algae
    c. Bacteria
    d. Earthworms

93. The Rhizobium is :
    a. Known to fix nitrogen in cereals
    b. A fungus symbiotically fixes nitrogen in legumes
    c. A bacteria found in roots of sugarcane
    d. Known to fix nitrogen in roots of legumes

94. Enzymes are more active at
    a. Low temperature
    b. Optimum temperature
    c. High temperature
    d. High humidity

95. Organism most tolerant to soil moisture stress is
    a. Fungi
    b. Protozoa
    c. Bacteria
    d. Actinomycetes

96. In leguminous crops, rhizobium enters root hair dissolving cell wall by
    a. Cellulase
    b. Chitinase
    c. Pectinase
    d. Lignolytic enzyme

97. Which of the following is P solubilising bacteria?
    a. Pseudomonas striata
    b. Nitrosomonas
    c. Thiobacillus ferroxidanse
    d. Fusarium oxysporum

98. Which group of soil microorganisms has highest biomass in soil
    a. Fungi
    b. Bacteria
    c. Algae
    d. Actinomycetes

99. Under what condition denitrification is a major mechanism of nitrogen loss from soil ?
    a. Well drained soil
    b. Fallow land
    c. Submerged soil
    d. Pasture land

100. The microorganism predominant in flooded rice soils is

a. Bacteria
b. Fungi
c. Actinomycetes
d. Protozoa

101. The microorganism that is considered most important as an indicator of soil quality is

a. Bacteria
b. Fungi
c. Actinomycetes
d. Algae

102. Mycorrhizae help in uptake of which nutrient

a. Phosphorus
b. Potassium
c. Nitrate
d. Phosphorus and Zinc

103. Nitrogen reduction in symbiotic systems takes place at

a. Nodules surface
b. Membrane envelop
c. Bacteroids
d. Cytochrome

104. Among the prominent greenhouse gases, the one whose emission is almost entirely dependent on soil related process are

a. Methane
b. Carbon dioxide
c. Chlorofluro carbon
d. Nitrous oxide

105. The 'Bangalore method' of composting is characterised by

a. Initial aerobic phase followed by anaerobic phase
b. Continuous aerobic phase
c. Initial anaerobic followed by aerobic phase
d. Continuous anaerobic phase

106. Lichens are the indicator of

a. SO_2 pollution
b. CO pollution
c. NOx pollution
d. CH_4 pollution

107. Net mineralization of soil nitrogen occurs when C : N ratio of organic matter is

a. $<20:1$
b. 20-30:1
c. $>30:1$
d. $>300:1$

108. Fraction of organic matter comparatively resistant to microbial degradation

   a. Protein
   b. Lignin
   c. Cellulose
   d. Hemicelluloses

109. Which organisms are microaerobic, free fixers or in association with grasses ?

   a. Azotobacler
   b. Azospirillum
   c. Rhizobium
   d. Frankia

110. Brady rhizobium sp. is suitable for

   a. Groundnut
   b. Soybean
   c. Cowpea
   d. Alfalfa

111. The function of leghaemoglobin found in the root nodules of legumes is

   a. Inhibition of nitrogenase activity
   b. Regulation of CO_2 supply to the nodule
   c. Regulation of O_2 supply to the nodule
   d. Inactivating the ATP

112. Molar masses of humus normally range between

   a. 20,000 - 1,00,000 g/mol
   b. 5,000 – 10,000 g/mol
   c. 500 - 1,000 g/mol
   d. 100 - 200 g/mol

113. The bioindicator organism for the faecal contamination of water is

   a. Clostridium sp
   b. Bacillus sp
   c. Algae
   d. E.coli

114. The enzyme responsible for the reduction of molecular nitrogen to the level of ammonia in the leguminous root nodule is

   a. Nitrogenase
   b. Nitrate reductase
   c. Nitrite reductase
   d. Carbonic anhydrase

115. Biofertilisers are :

   a. Products contain living cells of different types of micro-organism
   b. Mixture of vermicomposts and inorganic fertilisers

c. Mixture of vermicomposts and farm yard manures

d. Mixture of organic manures and inorganic fertilizers

116. Which one is essential for synthesis of auxin

a. Zinc
b. Sulphur
c. Potassium
d. Phosphorus

117.Choose the correct statement in given option

a. Nitrifying organisms are particularly sensitive to aeration.
b. Nitrite is oftenly found accumulated in soils.
c. The autotrophs have less efficient nitrifying system than most of the heterotrophs
d. Conversion of ammonia to nitrate is reductive process.

118. Choose the correct statement in given option

a. Both ammonium nitrogen and nitrate nitrogen can be immobilized by soil micro flora
b. Once ammonia is nitrified it is lost as volatilization
c. The mechanism of ammonium fixation is similar to that of phosphorus fixation
d. Ammonium fixation is increased by raising the levels of soil phosphorus

119. Two elements which are associated with biological nitrogen fixation are

a. Fe & Mo
b. Fe & Co
c. Mo & B
d. Co & Zn

120. Which one of the following micronutrients is a constituent of the enzymes carbonic anhydrase, Alcohol dehydrogenase and RNA polymerase

a. Iron
b. Iron
c. Manganese
d. Zinc

121. Phosphate solubilization from in organic sources by soil microorganisms is mainly through production of

a. Carbonic acid
b. Volatile fatty acids
c. Organic acids
d. Microbial enzymes

122. Which micronutrients manly influence the nitrogenase activity

a. B
b. Cu
c. Mn
d. Mo

123. Which group of microorganisms have highest biomass in soil

a. Bacteria
b. Actinomycetes
c. Fungi
d. Algae

124. The presence of Hartig net is a characteristic of

a. Endomycorrhizae
b. Ectomycorrhizae
c. Arbuscular mycorrhizae
d. Ericoid mycorrhizae

125. Name the enzyme responsible for reducing nitrogen (N_2) to ammonia (NH_3).

a. Peptase
b. Urease
c. Nitrogenase
d. Amylase

126. If azolla is grown in rice field along with rice crop, then how much nitrogen is fixed by azolla in rice field

a. 10 - 20 kg N/ha
b. 20 - 30 kg N/ha
c. 30 - 40 kg N/ha
d. More than 50 kg N/ha

127. Which legume crop fix highest N (kg N/ha/year)

a. Groundnut
b. Cow pea
c. Clover
d. Alfalfa

128. In India, the work on 'biogas' was first carried out by

a. J.N. Mukherjee
b. S.V. Desai
c. N.P.Datta
d. N.N. Goswami

129. Bacteria responsible for oxidizing $NO_2^- \rightarrow NO_3^-$ in soils

a. Nitrosomonas
b. Nitrobacter
c. Pseudomonas
d. Rhizobium

130. The process of conversion of organic form to inorganic form (mineral form) of nutrient element is called

a. Nitrification b. Denitrification

c. Mineralization d. Volatilization

131. Volume of soil under the influence of roots of growing plants is known as

a. Surface soil b. Sub-surface soil

c. Rhizosphere d. Solum

132. Rhizobium japonicum fixes N in symbiotic relationship with

a. Pea group b. Lupin group

c. Soybean group d. Phaseolus group

133. The processes of reduction of NO_3^- to N_2 or N_2O mediate by micro-organism is known as

a. Mineralization b. Nitrification

c. Denitrification d. Volatilization

134. The bacteria responsible for oxidizing $Fe^{2+} \rightarrow Fe^{3+}$ (ferrous to ferric form) under acidic environment

a. Pseudomonas b. Thiobacillus

c. Lactobacillus d. Orthobacter

135.Identify the microb which have also control the plant disease.

a. Thiobacillus b. Lactobacillus

c. Pseudomonas d. Orthobacter

136. The highest enzymatic activity is observed in which type of soil particle (clay size)

a. Clay 50 - 70 (μ) b. Clay < 50 (μ)

c. Clay 70 - 90 (μ) d. Clay < 90 (μ)

137. The micro-organism (bio-fertilizer) grows at the temperature 20 °C to 40 °C are called

a. Thermophile b. Mesophile

c. Psycrophile d. Hyper thermophile

138. Biogas is combination of following gases :

a. CH_4, H_2, O_2 and CO_2
b. C_2H_4, H_2, CO_2 and N_2
c. CO_2, H_2, O_2 and N_2
d. CH_4, CO_2, H_2 and N_2

139. Soil submergence in rice field leads to

a. Increase in pH of alkaline soil
b. Decrease in pH of acid soil
c. Increase in pH of acid soils
d. pH remains unaffected

140. The maximum permissible limit of BOD (mg/litre) of water for irrigation is

a. 100
b. 500
c. 1000
d. 2000

141. The term 'Spermosphere' refers to the zone around

a. Root
b. Germinating seed
c. Shoot
d. Bacterial cell

142. The plant component most resistant to microbial attack is

a. Fats
b. Starch
c. Protein
d. Lignin

143. A free living bacteria that fixes atmospheric nitrogen under aerobic conditions is

a. BGA
b. Azolla
c. Azotobacter
d. Rhizobium

144. Which one of the following microorganism is capable of improving the P availability in soil

a. Rhizobium
b. Azospirillum
c. Acetobacter
d. VAM

145. Microorganisms contributing maximum biomass of the soil is

a. Bacteria
b. Algae
c. Actinomycetes
d. Fungi

146. Who showed that legumes can obtain nitrogen from air

a. Robert Koch
b. J.B. Boussingault
c. Louis Pasteur
d. M. N. Bijerinck

147. Oxidation of S to SO_2^{2-} is strictly carried out under aerobic condition by the bacteria

a. Pseudomonas
b. Associative
c. Micrococcus
d. Thiobacillus sp.

148. Cyanobacterial are located near the surface of soil because they are..... in nature

a. Heterotrophic
b. Associative
c. Photoautotrophic
d. Chemoautotrophic

149. Critical C:N ratio of organic matter that dictates mineralization-immobilization of nitrogen is

a. 30:1
b. 25:1
c. 20:1
d. 10:1

150. Serial-dilution pour palte technique roughly enumerates........ of total microbes residing in soil

a. 1-10%
b. 10-20%
c. 20-30%
d. 50-60%

151. The process of decomposition of organic matter present in water by microorganisms using oxygen is known as

a. Nitrification
b. Symbiosis
c. Putrescibility
d. Osmosis

152. Symbiotic N-fixation without nodules which flourishes in low-land rice fields of tropical and sub tropical are as is called

a. Azolla-anabaena
b. Mycorrhizae-tree plant roots
c. Rizobium-legume
d. Azotobacter-non legume

153. In the organic manures, the organic compounds vary greatly in their rate of decomposition. Among the tested four constituents, which has the slowest rate of decomposition is

a. Sugars
b. Lignin
c. Celluloses
d. Hemicelluloses

154. Active root zone of the growing plants is designated as

a. Rhizosphere
b. Rhizoctonia
c. Rhizobium
d. Mycorrhzae

155. The association between to organisms where a partner deriver benefits of the other partner without affecting is called

a. Proto-corporation
b. Ammensalism
c. Commensalism
d. Neutralism

156. Which of the following microbial associations involves algae as one partner of this association

a. Mycorrhizae
b. Lichen
c. Reindeer mosses
d. Predation of Bdellovibrio spp.

157. Plant growth promoting rhizobacteria (PGPR) control deleterious rhizobacteria (DRB) by arresting the production of

a. HCN
b. HCl
c. H_2S
d. Siderphore

158. Mycorrhiza-infested plants show better ability of absorption of the nutrient, especially

a. Iron in high fertility soils
b. Potassium in desert soils
c. Manganese in marshy soils
d. Phosphorus in low fertility soils

159. The largest number of species of soil bacteria belong to

a. Azotobacter
b. Pseudomonas
c. Aerobacter
d. Bacillus

160. The approximate molecular weight of the protein not containing Mo in nitrogenase enzyme is

a. 60,000
b. 1,00,000
c. 80,000
d. 2,00,000

161. Among the following, the correct sequence of biotic biopolymer degradation pathway is

a. Recalcitrant biopolymers - a humin acid a humic acid - a fulvic acid
b. Fulvic acid - a humic acid - a recalcitrant biopolymers - a humin acid
c. Humin acid - a humic acid - a fulvic acid - a recalcitrant biopolymers
d. Humic acid - fulvic acid - a recalcitrant biopolymers - a humin acid

162. Ammonifying micro-organisms are most active in the pH range of

a. 8.5-9.5
b. 5.5-6.5
c. 7.5-8.5
d. 4.5-5.5

163. With increase in oxygen content of the soil, the loss of nitrogen through denitrification

a. Increases
b. Decreases
c. Is not affected
d. May increase or decrease

164. In the process of denitrification, the organisms obtain

a. Carbon for cell synthesis
b. Energy for cell synthesis
c. Ammonia for assimilation
d. Nitrates for assimilation

165. The nutrient which imparts some resistance to plants against fungal and bacterial diseases is

a. Nitrogen
b. Potassium
c. Phosphorus
d. Sulphur

166. Which organism is leaf-inhabitant

a. Azotobacter
b. Aerobacter
c. Beijernekia
d. Clostridium

167. The major organic cementing agent in soil aggregate formation is

a. Lipids
b. Proteins and protein derivatives
c. Polysaccharides
d. Organic acids

168. If a soil solution is concentrated four times, the K-Ca activity ratio in solution will be come

a. Half
b. One-fourth
c. Double
d. One-eighth

169. Coralloid roots of Cycas have symbiotic association with

a. Actinomycetes
b. Bacteria
c. B.G.A.
d. Fungi

170. Concentration of nitrogen in soil air is

a. 70 %
b. 74 %
c. 78 %
d. 82 %

171. C:S ratio of organic residues for sulphur mineralization is

a. <200
b. 400-500
c. 200-400
d. >500

172. Cunninghamella test is done to assess the organism for its

a. N_2-fixing capacity
b. S-oxidizing capacity
c. P-solubilising capacity
d. Fe-oxidizing capacity

173. On an average, use of 10 ton/ha of Azolla can supply N equivalent to

a. <10 kg/ha
b. 25-30 kg/ha
c. 60-70 kg/ha
d. >100 kg/ha

174. The most important non-nodule nitrogen fixing system is

a. Azolla-Anabaena
b. Blue green algae
c. Rhizobium-legume
d. Azotobacter

175.Maximum nitrification takes place at

a. 10% O_2
b. 20% O_2
c. 30% O_2
d. 40% O_2

176. Hartig nets in soil produce by

a. Rhizobium
b. Ectomycorrhiza
c. Endomycorrhiza
d. None

177. Presence of 'Hartig net' is a characteristic of

a. Endomycorrhiza
b. Ectomycorrhiza
c. Arbuscular mycorrhiza
d. Ericoid mycorrhiza

178. Rhizosphere is dominated by

a. Gram negative spore-forming bacteria
b. Gram negative non-spore forming bacteria
c. Gram positive spore-forming bacteria
d. Gram positive non-spore forming bacteria

179. Which of the following is the typical thermophillic fungal species which develops during composting

a. Bacillus
b. Streptothermophilus
c. Humicolla
d. Bacillus and Strepothermophilus

180. End product of Dinitrification process is

a. N_2
b. NO_3
c. N_2O and N_2
d. NH_3

181. Nitrosomonas and Nitrobacter cause nitrification are an obligate

a. Autotrophic bacteria
b. Heterotrophic bacteria
c. Both (a) and (b)
d. Saprophytes

182. Ammonia volatilization from composting can be checked by the application of

a. Vermiculite
b. Gibsite
c. Kaolinite
d. Montmorillonite

183. Which organism fixes atmospheric nitrogen symbiotically

a. Rhizobium
b. Nitrobacter
c. Nitrosomonas
d. Nitrosococcus

184. Among the following GHG's, which has the highest global warming Potential

a. CO_2
b. CH_4
c. N_2O
d. H_2O

185. What is the fraction of soil organic matter which is soluble both in alkali and acid ?

a. Humic acid
b. Fulvic acid
c. Hymatomelonic acid
d. Humin acid

186. For nitrification, the optimum pH is

a. 6.5
b. 7.5
c. 8.5
d. 9.5

187. Which one is associative N-fixing bacteria ?

a. Rhizobium
b. Azospirilluin
c. Azotobacter
d. Acetobacter

188. Which of the given below is/are facultatively anaerobic bacteria ?

a. Azotobacter
b. Bacillus
c. Klesiella
d. Both (b) and (c)

189. Azotobacter is a

a. Symbiotic bacteria
b. Free living bacteria
c. Associative bacteria
d. Free living and associative bacteria

190. Carbon becomes available to crop plant in the form of

a. Amino acids
b. Carbon dioxide
c. Carbonates
d. Elemental carbon

191. Which gas is maximum in biogas

a. CH_4
b. H_2
c. CO_2
d. N_2

192. The process of nutrient enrichment of water bodies leading to detrimental change in the water quality is called

a. Eutrophication
b. Oligotrophication
c. Putrification
d. Contamination

193. Which one of the following fractions of organic matter is comparatively resistant to microbial degradation

a. Cellulose
b. Lignin
c. Protein
d. Hemicellulose

194. Which element is the metal component of urease enzyme ?

a. Silicon
b. Sodium
c. Molybdenum
d. Nickel

195. Molybdenum is a component of

a. Urease
b. Nitrogenase
c. Nitrate reductase
d. Both (b) and (c)

196. Which one of the following biofertilizer helps in phosphorus nutrition

a. Azotobacter sp.
b. Rhizobium japonicum
c. Nostoc sp.
d. Bacillus megaterium

197. Fe is an important component of enzymes

a. Nitrogenase
b. Nitrate reductase
c. Nitrite reductase
d. All of these

198. Dehydrogenase activity is a good index of biological activity of soil because it plays role in

a. Carbon metabolism
b. Respiration
c. Synthesis of cell macromoleculesd. Cell division

199. Association of two microorganisms in which, one is benefitted while otherremains unaffected is known as

a. Commensalism
b. Protocooperation
c. Ammensalism
d. Competition

200. Reduction reaction (NO_2 to NH_3) for nitrate assimilation by enzyme nitrite reductase takes' place in

a. Protoplasm
b. Cytoplasm
c. Chloroplast
d. Nucleus

201. Aspergillus awamori is a

a. K-mobilizer
b. Organic matter decomposer
c. S-oxidizer,
d. P-solubilizer

202. Who isolated Bacillus radicicola, the organism responsible for N fixation in 1888

a. S Winogradsky
b. MW Beijerinck
c. Milton Whitney
d. JB Lawe's

203.'cfu' is a unit for measuring

a. soil pollution
b. Metal contamination
c. Soil quality
d. Microbial population

204. Good earth worm activity is measure of

1. Soil nitrogen-supplying potential
2. Microbial activity
3. Biological soil health
4. Physical soil health

Select the correct answer using the codes:

a. (1), (2) and (3)
b. (2), (3) and (4)
c. (3) and (4)
d. (1) and (2)

205. Which of the following are nitrification inhibitors

1. Potassium azide
2. Dicyandiamide
3. Acetohydroxamate
4. Thiocarbamate

Select the correct answer using the codes:

a. (1) and (2)
b. (3) and (4)
c. (2) and (3)
d. (1), (2) and (4)

206. Which of the following are denitrifying bacteria

1. Micrococcus denitrificans
2. Pseudomonas aeruginosa
3. Thiobacillus thioparas
4. Nitrosospira

Select the correct answer using the codes:

a. (1), (2) and (4)
b. (1), (2) and (3)
c. (2) and (3)
d. (3) and (4)

207. Match List I with List II

a. Saw dust	1. 8 : 1
b. Soil microorganisms	2. 10 :1
c. Soil organic matter	3. 1000 :1
d. Farm yard manure	4. 20 : 1

208. Match List I (Mineralization/immobilization) with List II (C/P ratio)

a. Net immobilization of inorganic P	1. < 200
b. Net mineralization of organic P	2. 200 -300
c. Net immobilization of organic P	3. >300
d. No gain/loss of inorganic P	4. >400

209. Match List I (Microbial process) with List II (Micro-organism)

a. Nitrification	1. Thiobacillus
b. Denitrification	2. Clostridium, Azotobacter
c. Non-symbiotic nitrogen fixation	3. Pseudomonas, Bacillus
d. Symbiotic nitrogen fixation	4. Nitrosomonas, Nitrobacter

210. Match List I (Micro - organisms) with List II (Functions)

a. Thiobacillus	1. Conversion of NH_2-N to NO_3- N
b. Clostridium	2. Conversion of NH_4-N to NO_2- N
c. Nitrosomonas	3. Asymbiotic fixation of nitrogen
d. Nitrobacter	4. Hydrolysis of urea

211. Match List I with List II

	List I		List II
a.	Phospholipid fatty acids	1.	Phosphatase
b.	Phenyl mercuric acetate	2.	Dehydrogenase
c.	Fumigation extraction	3.	Biodiversity
d.	Tri-phenyl tetrazolium chloride	4.	Microbial biomass-C
e.	P-nitrophenol	5.	Urease inhibitor

212. Match List I with List II

	List I		List II
a.	Photoautrophs	1.	Energy from sunlight, C from CO_2
b.	Chemoautotrophs	2.	Exist as free living organisms but fix N_2
c.	Non-symbiotic N_2 fixers	3.	Energy from sunlight, C from organic matter
d.	Anaerobes	4.	Can use chemical species other than free oxygen as electron acceptors
e.	Photoheterotrophs	5.	Energy from oxidation of inorganic substances, C from CO_2

213. Match List I with List II

	List I		List II
a.	Legume-Rhizobium symbiosis	1.	Sesbania rostrata
b.	Non-legume Frankia symbiosis	2.	Rice field
c.	Azolla-Anabaena symbiosis	3.	Fish
d.	Non-specific associative N_2 fixers	4.	Alnus, Myrica
e.	Phytoplankton	5.	Azotobacter, Azospirillum

214. Match List I with List II

	List I		List II
a.	P solubilizationa	1.	Gigaspora
b.	S oxidation	2.	Thiobacillus
c.	Vermi compost	3.	Pseudomonas
d.	VAM	4.	Azotobacter
e.	Diazotrophe	5.	Eisenia

Answers (Soil Microbiology)									
1.a	2.a	3.a	4.b	5.a	6.a	7.a	8.d	9.c	10.b
11.a	12.a	13.b	14.c	15.d	16.d	17.a	18.c	19.b	20.c
21.b	22.c	23.a	24.a	25.a	26.a	27.c	28.c	29.b	30.c
31.a	32.b	33.a	34.a	35.b	36.b	37.a	38.d	39.b	40.c
41.c	42.a	43.a	44.c	45.b	46.d	47.c	48.b	49.a	50.d
51.a	52.d	53.b	54.c	55.c	56.b	57.d	58.d	59.d	60.d
61.b	62.b	63.b	64.c	65.b	66.d	67.d	68.d	69.b	70.c
71.b	72.a	73.b	74.b	75.a	76.b	77.b	78.b	79.d	80.a
81.c	82.d	83.a	84.b	85.a	86.d	87.d	88.c	89.c	90.c
91.b	92.c	93.d	94.b	95.d	96.d	97.a	98.a	99.c	100.a
101.a	102.a	103.c	104.a	105.a	106.a	107.a	108.b	109.b	110.b
111.c	112.a	113.d	114.a	115.a	116.a	117.a	118.a	119.a	120.d
121.c	122.d	123.c	124.b	125.c	126.c	127.d	128.b	129.b	130.c
131.c	132.c	133.c	134.b	135.c	136.b	137.b	138.d	139.c	140.a
141.b	142.d	143.c	144.d	145.d	146.b	147.d	148.c	149.b	150.a
151.c	152.a	153.b	154.a	155.c	156.b	157.d	158.d	159.d	160.a
161.b	162.c	163.b	164.a	165.b	166.c	167.c	168.c	169.c	170.c
171.a	172.c	173.b	174.d	175.b	176.b	177.b	178.b	179.c	180.c
181.a	182.b	183.a	184.c	185.b	186.c	187.b	188.d	189.b	190.b
191.a	192.a	193.b	194.d	195.d	196.d	197.d	198.b	199.a	200.c
201.d	202.b	203.d	204.b	205.a	206.b				
207. a.3 b.1 c.2 d.4					208. a.4 b.1 c.3 d.2				
209. a.4 b.3 c.2 d.1					210. a.3 b.4 c.2 d.1				
211. a.3 b.5 c.4 d.2 e.1					212. a.1 b.5 c.2 d.4 e.3				
213. a.1 b.4 c.2 d.5 e.3					214. a.3 b.2 c.5 d.1 e.4				

4

Soil Chemistry

1. Central Fertilizer Quality Control & Training Institute (CFQC&TI) is situated at

   a. Mysore b. Karnal

   c. Faridabad d. Coimbatore

2. The pH of acid sulfate soil is less than 3.5. In India it is found in _________ state.

   a. Odisha b. Gujrat

   c. Kerala d. West Bengal

3. Ionic potential is maximum in case of

   a. Cs b. Li

   c. K d. Na

4. Who is the father of soil chemistry?

   a. J.V. Liebig b. V.V. Dokuchaev

   c. Helter d. Stanley

5. In humid region generally

   a. Acid soils are formed b. Alkali Soils are formed

   c. No soil formation takes place d. All of these

6. Flame photometer is used in the determination of :

   a. P b. K

   c. S d. Zn

7. Layer silicates are

   a. Phyllosilicates b. Sesquioxide clays

   c. Allophanes d. All above

8. Akiochi disease of rice is caused by _________

a. Hydrogen cyanide (HCN)
b. Hydrogen peroxide (H_2O_2)
c. Hydrogen sulphide (H_2S)
d. Hydrogen isocyanide (HNC)

9. Silica tetrahedron had

a. Four sided arrangement
b. Six sided arrangement
c. Eight sided arrangement
d. All above

10. Montmorillonite is

a. Expanding clay mineral
b. Limited expanding mineral
c. Non expanding mineral
d. None of the above

11 .In the octahedral sheet of a crystal unit of silicate clays, Al octahedral are bound together by

a. Shared oxygen atoms
b. OH-H bond
c. H-bond
d. OH-bond

12. Choose the incorrect statement

a. Liming of acid soils decrease phosphate availability
b. Phosphate maximum available to pH of 6.5 to 7.5
c. Reduction takes place under mostly anaerobic conditions
d. Claying of soil increase sulphur mineralization

13. Unit of CEC is

a. Cmol (P+)/kg
b. meq 100/g
c. Moles/lit
d. Both (a) and (b)

14. Active acidity is always greater

a. True
b. False
c. Not always true
d. None of above

15. Chemical formula of potassium feldspar is

a. $KAlSi_5O_8$
b. $KAlSiO_3$
c. $KAlSi_3O_8$
d. $KAlSi_3O_6$

16. The degree of acidity increases with

a. OH- ion decrease
b. OH- ions remain stable
c. OH- ions increases
d. All above

17. The pH scale is given by

a. Simpson
b. Soloman
c. Sorenson
d. Stuwart

18. During decomposition of plant materials, the protein are hydrolysed in to

a. Polypeptides
b. Humin
c. Quinines
d. Humic acid

19. Availability of which micronutrient increases with acidity

a. Fe
b. Mn
c. Zn
d. All of these

20. Principal mineral constituent of rock phosphate is

a. Crandallite
b. Apatite
c. Phlogopite
d. Struvite

21. If a soil solution is concentrated four times the K-Ca activity ratio in solution will

a. Decrease two times
b. Increase two times
c. Decrease sixteen times
d. Increase sixteen times

22. Most soluble Fe-minerals in soil is

a. y-Fe_20_3
b. FeOOH
c. $Fe(OH)_3$
d. α-Fe_20_3

23. Initially, micas lose their structural K^+ rapidly in the weathering solution through which of the following?

a. Cation exchange
b. Dissolution
c. Hydrolysis
d. Desorption

24. The ion with highest ionic potential is

a. SO_4^{2-}
b. Na^+
c. Al^{3+}
d. K^+

25. The typical peaty saline highly acidic (pH 3.5) soil is also known as

a. Kari soil b. Chalka soil

c. Pokkali soil d. Regur soil

26. Which of the following formula of hematite

a. Fe_2O_3 b. Fe_3O_4

c. FeOOH d. Fe_3OH

27. The presence of Vermiculite in soil indicates the potential problem of

a. K fixation b. P fixation

c. Acidity d. Ammonia volatilization

28. PPM is equal to

a. 0.0001% b. µg/g

c. mg/kg d. All of these

29. Langmuir isotherm often describes successfully which one of the following?

a. Chemisorption b. Physical sorption

c. Multilayer adsorption d. Multilayer physical adsorption

30. Which of the following is the specific surface (m^2/g) of mica mineral group

a. 700 - 800 b. 100 - 120

c. 5 - 20 d. > 800

31. Choose the incorrect statement

a. Soil acidity is caused by ionizable hydrogen ion or protons

b. Base rocks highly weathered as compared to acid rocks

c. Acidic soil less fix phosphorus as compared to alkaline soil

d. Blocky structure common in humid regions

32. CEC of which clay mineral is highest

a. Kaolinite b. Montmorillonite

c. Vermiculite d. Illite

33.Micas are examples of

a. Tectosilicates
b. Orthosilicates
c. Phyllosilicates
d. Ionosilicates

34. Soil silicate clay mineral, Illite is responsible for fixation of _________ from applied fertilizer

a. PO_4^- and K^+
b. NH_4^+ and K^+
c. SO_4^- and K^+
d. NO_3^- and K^+

35. Dickite belong to which group of mineral

a. Kaolinite
b. Mica
c. Smectite
d. Amorphous

36. The pH of the soil solution determines

a. Active acidity
b. Exchange acidity
c. Total acidity
d. None

37. In the 1:1 layer silicate, which of the following of tetrahedral sheet is being shared with the octahedral sheet

a. Apical oxide (O^{2-}) ion
b. OH^-
c. Si^{++}
d. Al^{3+}

38. The mechanisms of ammonium and potassium fixation are similar due to their almost same

a. Ionic size
b. Polarizabilites
c. Both (a) & (b)
d. Solubility

39. In which of the following 1:1 layer silicate mineral, 1:1 layer are separated by a layer of H_2O molecules which is fully hydrated

a. Kaolinite
b. Halloysite
c. Anauxite
d. Dickite

40. Micas may have isomorphous substitution

a. Predominantly in tetrahedral layer
b. Predominantly in octahedral layer
c. In both the layers equally
d. In octahedral layer only

41. The permanent negative charges on silicate clay are due to
    a. Isomorphus substitution by higher valence cations for lower valency cations
    b. Isomorphus substitution by lower valency cations for higher valency cations
    c. Dissociation of H ions at broken edge
    d. All the these
42. Buffering capacity of soil influenced by
    a. Amount of clay colloids
    b. Amount of organic colloids
    c. Type of clay colloids
    d. All of the above
43. In a octahedral sheet of layer silicate if the magnesium atoms are coordinated with six oxygen hydroxyls then the sheet is called as
    a. Dioctahedral
    b. Eight sided octahedron
    c. Trioctahedral
    d. None of the above
44. The concentration of hydrogen ion in distilled water at 25°C is
    a. 10^{-4} mols
    b. 10^{-5} mols
    c. 10^{-6} mols
    d. 10^{-7} mols
45. Extraction of P by Brays method is done by
    a. $NaHCO_3$
    b. NH_4F
    c. NH_4Cl
    d. Na_2CO_3
46. In a octahedral sheet of layer silicate if the Al^{+3} atoms are co-ordinated with six oxygen /hydroxyls then the sheet is called as
    a. Dioctahedral
    b. Eight sided octahedron
    c. Trioctahedral
    d. None of the above
47. In order to impart a net negative charge to clay micelle the isomorphous substitution takes place between ions having
    a. Different Valency but more or less similar size
    b. Similar valency and more less similar size
    c. Similar valency but different
    d. Different valency as well as different size

48. Carbonates & bicarbonates present in irrigation water are determined by titration with standard

a. NaOH
b. HCl
c. H_2SO_4
d. HNO_3

49. The sequence of anion exchange will be

a. < Sulfate < Nitrate < Phosphate < Nitrite
b. > Sulfate > Nitrate > Phosphate > Nitrite
c. < Nitrite < Phosphate < Nitrate < Sulfate
d. > Nitrite > Phosphate > Nitrate > Sulfate

50. The soils having EC < 4 dS/m, ESP > 15 and pH more than > 8.5 the soils is

a. Saline
b. Alkali
c. Saline alkali
d. Degraded alkali

51. Tri acid digestion is carried out by using a mixture of

a. HCl : NaOH: H_2SO_4
b. H_2SO_4: HCl : $CHIO_4$
c. HNO_3 : H_2SO_4 : $HClO_4$
d. NaOH : HNO_3 H_2SO_4

52. One ml of 1 N H_2SO_4 is equiralent to

a. 0.0014 g N
b. 0.014 g N
c. 0.14 g N
d. 1.4 g N

53. The $CaCO_3$ is commonly absent in the soils having pH

a. < 7.0
b. > 7.0
c. < 8.4
d. > 8.4

54. Microbial conversion of nitrate into chemical nitrogen is called

a. Denitrification
b. Nitrate reduction
c. Nitrogen fixation
d. None of these

55. Soil testing programme in India was initiated in

a. 1947
b. 1952
c. 1954
d. 1964

56. Which of the following layer silicate mineral fixes more `K.' in soil clay complex

a. Illite
b. Vermiculite
c. Kaolinite
d. Montmorillonite

57. Soil pH is measured by making soil:water suspension in the ratio of

a. 1:10
b. 1:2.5
c. 1:5
d. 1:1

58. Lime potential of soil can be written as

a. p (Ca+Mg)-pH
b. pH-p (Ca+Mg)
c. 1/2 p (Ca+Mg)-pH
d. pH -1/2 p (Ca+Mg)

59. In case of mica, excess of negative charge in tetrahedral layer is balance by

a. K^+
b. Al^{3+}
c. Mg^{2+}
d. Fe^{3+}

60. High amount of K_2O (Potash) in the soil indicate the presence of

a. Illite
b. Vermiculate
c. Chlorite
d. Montmorillonite

61. The factor responsible for formation of acid soil is

a. Low rainfall
b. High rainfall
c. High evaporation
d. Low rainfall and high evaporation

62. How much is the layer charge per half cell of vermiculite

a. 0.5-0.6
b. 0.6-0.9
c. < 0.25
d. < 0.6

63. What is the coordination number of Al^{3+} as central cation in phyllosilicate structure

a. 4
b. 10
c. 6
d. 8

64. Laboratory method of sulphur identification in soil is

a. Olsen's method
b. Walkley and black method
c. Turbidity method
d. Alkaline permanganate method

65. Carboxylic group of soil humic substances ionizes at what pH

a. 5.5-6.0
b. 9.0
c. > 9.0
d. 12.0

66. 1 ml of 0.1 M sulphuric acid is equivalent to

a. 0.028 g of N
b. 0.014 g of N
c. 0.0028 g of N
d. 0.0014 g of N

67. The chemical formula of magnetite (hydrous oxide) is

a. Fe_2O_3
b. Fe_3O_4
c. Fe_3O_3
d. Fe_2O_4

68. Eh of distilled water at pH 7 should be

a. 812 mV
b. 612 mV
c. 712 mV
d. 512 mV

69. A decrease of one pH unit is accompanied by an increase in Eh of

a. 58 mV
b. 5.8 mV
c. 0.58 mV
d. 0.058 mV

70. Who developed method for estimation of lime requirement

a. Sckooner (1952)
b. Shoemaker et al. (1961)
c. Stanford and English (1949)
d. Olsen et al. (1954)

71. The ZPC of the gibbsite is

a. < 9
b. > 9
c. 5
d. 7

72. Zinc containing mineral is

a. Pyrolusite
b. Olivine
c. Haematite
d. Willemite

73. Tannery and textile industrial effluents are characterized by

a. High BOD, COD
b. Low BOD. COD
c. Medium BOD, COD
d. High COD

74. Black cotton soil is dominated by which type clay minerals?

a. Kaolinite
b. Montmorillonite
c. Chlorite
d. Vermiculite

75. The 2 : 2 type of mineral is the characteristics of

a. Illite
b. Montmorillonite
c. Chlorite
d. Vermiculite

76. Choose the false statement in given options.

a. The availability of Mo decreases as soil pH decreases.
b. Cuprous ions are more soluble than cupric ions, when soil pH < 7.
c. Overliming of soil may induce Zn deficiency and decrease Zn availability.
d. The availability of B, Cu, Fe, Mn, and Zn decreases usually soil pH decreases.

77. Blue baby syndrome is caused by excesses presence of

a. Cd
b. NO_3
c. Pb
d. Ni

78. What will be the effect of addition of electrolyte on zeta potential of a soil colloidal system

a. Increase
b. Decrease
c. No change
d. Initially increased than decrease

79. In electrophoresis, which of the following does move

a. Clay particles
b. Solution
c. Both clay particles and solution
d. Ions

80. For the cations of varying valencies, the lyotropic series in respect of absorbability can be given by

a. $Mg^{2+} > Ca^{2+} > Al^{3+} > Na^{+}$
b. $Ca^{2+} > Mg^{2+} > Na^{+} > Al^{3+}$
c. $Ca^{2+} > Al^{3+} > Mg^{2+} > Na^{+}$
d. $Al^{3+} > Ca^{2+} > Mg^{2+} > Na^{+}$

81. Which of the following is not the assumption of the theory of diffuse double layers

    a. The charged particle has flat surface without edge effect.
    b. The exchange sites are uniformly distributed over the entire planer surface.
    c. The ions act as point charge.
    d. The interaction between ion is extensive

82. In a clay-water system, on increasing the electrolyte concentration four times, the thickness of the diffuse double layer reduces to

    a. 1/2 b. 1/3
    c. 1/5 d. 1/4

83. What is the correct form of Henderson/Hasselbaich equation which is used to compute the pH of a buffer mixture of a week acid and its salt

    a. $pH = pKa + \log (C_{acid}/C_{salt})$ b. $pH = pKa + (C_{salt}/C_{acid})$
    c. $pH = pKa + \log (C_{sait}/C_{acid})$ d. $pH = pKa + P (C_{sait}/C_{acid})$

84. Which of the following is responsible for the green colour at the end of the titration in organic carbon determination

    a. Cr^{6+} b. Cr^{3+}
    c. C^{4+} d. C^{4-}

85. Which of the following is not result of eutrophication

    a. Accumulation of algae
    b. High aquatic plant/fish mortality
    c. Detrimental in water change quality
    d. Improving the water quality and good eco system.

86. Which is not related with podzolization

    a. Removal of Fe and Al oxides b. Acid soil
    c. Tropical climate d. High rainfall

87. Saline soils have the characteristics of

    a. EC > 4 mmhos/cm, pH > 8.5 and ESP > 15
    b. EC < 4 mmhos/cm, pH > 8.5 and ESP > 15

c. EC > 4 mmhos/cm, pH < 8.5 and ESP < 15

d. EC < 4 mmhos/cm, pH < 8.5 and ESP < 15

88. The catastrophic explosion at Beirut port of Lebanon, on 4 August 2020 caused by

a. Calcium nitrate
b. Ammonium nitrate
c. Calcium cyanamide
d. Ammonium phosphate

89. Illuviation is the process of

a. Accumulation
b. Leaching
c. Weathering
d. Transformation

90. Montmorillonite is....... minerals

a. Dimorphic
b. Tetramorphic
c. Hydromorphic
d. Trimorphic

91. Effective cation exchange capacity of soil is measured at

a. Neutral pH
b. Acidic pH
c. Alkaline pH
d. Natural pH of the soil

92. In determination of Ca+Mg by EDTA which of the following indicator is used in titrating Ca alone

a. Erichrome Black T
b. Phenolphthalein
c. Diphenylamine
d. Calcon

93. Green colour in flame test shows the presence of

a. Potassium
b. Calcium
c. Lithium
d. Sodium

94. Layer silicate structure cannot be formed without

a. Aluminium
b. Nitrogen
c. Potassium
d. Zinc

95. In the colorimetric determination of Fe, which of the following reagent is used to reduce from ferric to ferrous form

a. o-phenanthroline
b. Hydroxylamine hydrochloride
c. Ascorbic acid
d. $SnCl_2$

96. During the extraction of micronutrient cations with DTPA (Norvell and Lindsay) how much TEA is protonated

a. 50%   b. 25%

c. 75%   d. 100%

97. What is the critical limit of deficiency of hot water extractable B (PPM) in soil

a. 0.5   b. 1.0

c. 1.5   d. 0.1

98. Which one of the following is used as extratant for available P estimation in Olsen's method

a. 0.5 M $NaHCO_3$ (pH:7.5)   b. 0.5 M $NaHCO_3$ (pH:8.5)

c. 0.5 M $NaHCO_3$ (pH:5.5)   d. 0.5 M $NaHCO_3$ (pH:6.5)

99. The equivalent weight of H_2SO_4 is

a. 49   b. 36

c. 98   d. 14

100. The pH of the acid sulfate soils occur as

a. 5-6   b. 6-7

c. 4-5   d. <4

101. Adsorption of sulphate by soils is negligible at pH value of

a. >6.5   b. >3.5

c. <6.5   d. >5.0

102. In determination of P, phosphate-ammonium molybdate does undergo which of the following process to produce blue colour

a. Reduction   b. Oxidation

c. Solubilisation   d. Precipitation

103. The value of ESP in alkali soils is always

a. Greater than 15   b. Less than 15

c. Any value   d. None

104. Which of the following reagent is used to extract mineral N from moist soil

a. NaCl
b. MgCl
c. KCl
d. NH_4Cl

105. In Kjeldahl digestion, soil to convert organic N to ammonium N which of the following is used as catalyst

a. Na_2SO_4
b. K_2SO_4
c. $CuSO_4$
d. $CaSO_4$

106. The H^+ associated with the $(OH)^-$ group of inorganic colloids contributes to

a. pH dependent acidity
b. Permanent charge acidity
c. Active charge acidity
d. Total charge acidity

107. The Zn mineral which ultimately control Zn solubility in soils is

a. Franklinite
b. Zincite
c. Willemite
d. None of these

108. Classification of irrigation water as determined by a diagram which represents the parameters of

a. EC and SAR
b. pH and EC
c. pH and salt content
d. EC and salt content

109. Sources of negative charges on silicate clays is/are

a. Exposed crystal edges
b. Isomorphus substitution
c. Anion exchange
d. All of these

110. pH of the solution containing 0.0001g H^+ per litre will be

a. 2
b. 3
c. 4
d. 5

111. Phenolic group of soil humic substances ionizes at what pH

a. 9.0
b. 5.5 - 6.0
c. 6.0
d. 5.5

112. The pH of a calcareous soil in equilibrium with atmospheric CO_2 is

a. 6.8
b. 7.5
c. 8.2
d. 8.5

113. A soil which has pH < 8.5, ESP < 15 and EC > 4 dS/m at 25°C is called

a. Saline soil
b. Alkaline soil
c. Saline - alkaline soil
d. None of these

114. A chemical bond formed between two atoms by sharing an electron is

a. Ionic bond
b. Hydrogen bond
c. Covalent bond
d. Vander wall bond

115. If the nitrate formation in a soil wear independent of the amount of ammonical-N, in that case the reaction would be considered as

a. Zero order
b. First order
c. Second order
d. Either first or second order

116. Which part of spectrophotometer converts light in to electric energy

a. Filter
b. Photo cell
c. Prism
d. Cuvette

117. Under submerged conditions, green manuring in transplanted rice result in:

a. Increase in redox potential
b. Increase in pH
c. Decrease in redox potential
d. Decrease in ionic strength

118. Formula for SAR is

a. $SAR=Na^+/(Ca^{2+}+Mg^{2+})^{1/2}$
b. $SAR=Na^+/\{(Ca^{2+}+Mg^{2+})/2\}^{1/2}$
c. $SAR= (Na^+/(Ca+Mg)/2)$
d. $SAR=Na^+/(Ca^{2+}+Mg^{2+})$

119. Yellow colour in flame test shows the presence of

a. Potassium
b. Calcium
c. Lithium
d. Sodium

120. Iron and aluminium oxide clays carry negative charge in soils at

a. High pH
b. Low pH
c. Neutral pH
d. Near neutral pH

121. Which of the following is determined by the use of Walkley and Black's method

   a. Potassium
   b. Sulfur
   c. Organic matter
   d. Cation exchange capacity

122. Total nitrogen in soil is determined by

   a. Alkaline permanganate method
   b. Olsen's method
   c. Kjeldahl process
   d. Bray's method

123. Magnesium in octahedral sheet of clay mineral indicates

   a. Dioctahedral
   b. Trioctahedral
   c. Tetrahedral
   d. Monohedral

124. Anion exchange capacity is much greater in soils with high

   a. Kaolinite
   b. Vermiculite
   c. Montmorillonite
   d. Illite

125. Which of the following is 2:1 expanding type clay mineral

   a. Illite
   b. Vermiculite
   c. Kaolinite
   d. Chlorite

126. Which instrumental method is used for estimation of micronutrients and heavy metals

   a. Atomic absorption spectrophotometer
   b. Flame photometer
   c. Spectrophotometer
   d. Polarimeter

127. Gapon's equation is widely used equation to describe

   a. P status of soils
   b. CEC of soils
   c. Organic matter of soils
   d. Water status of soils

128. Emission intensity in flame emission spectroscopy depends upon

   a. Number of excited atoms
   b. Temperature
   c. Number of excited atoms & temperature
   d. Number of unexcited atoms

129. Which is not suitable for podzolization

a. Acid soil
b. High organic matter
c. Hot and humid climate
d. Abundant moisture

130. Silica : sesquioxide (R_2O_3) ratio for tropical ecoregion is

a. More than 3.5
b. 3 to 3.5
c. 2 to 3
d. Less than 2

131. Which mineral is an important source of potassium

a. Pyroxenes
b. Albite
c. Plagioclase
d. Feldspars

132.Which one in phyllosilicates in nature

a. Mica
b. Quartz
c. Feldspars
d. Olivine

133. Which is the principal buffer mechanism in soil

a. Organic carbon
b. Calcium carbonate
c. Cation exchange capacity
d. Soil structure

134. Iron pyrites are found in which soils

a. Acid soils
b. Alkaline soils
c. Acid sulphate soils
d. Saline alkali soils

135. Dominance of which increases the thickness of double diffuse layer

a. Monovalent cations
b. Divalent cations
c. Trivalent cations
d. Pentavalent cations

136. Which of the following processes explains the formation of secondary crystalline colloids from the mica and feldspars present in the parent rock

a. Hydrolysis
b. Solution
c. Coagulation
d. Carbonation

137.The crystal units of montmorillonite are held together by

a. O-O linkages
b. O-H linkages
c. H bonding
d. K^+ ions

138. Which method is applied for testing of Micronutrients and Heavy metals

a. EDDHA
b. HEDTA
c. Oxalic acid
d. DTPA

139. Which element imparts red colour to the soil

a. Ca
b. Fe
c. S
d. Mg

140. What common extractants are used for the determination of exchangeable cations in soils

a. HCl
b. NH_4OH
c. NH_4Cl
d. NH_4OAc

141. Sulphur in soil is determined by using

a. $CaCl_2$
b. KCl
c. NH_4OAC
d. $NaCO_3$

142. The most electronegative element in the periodic table is

a. Chlorine
b. Fluorine
c. Gold
d. Silver

143. Devarda alloy is composed of

a. 50% Cu, 45% Al and 5%Zn
b. 50% Al, 45%Cu and 5%Zn
c. 50% Zn, 45%Al and 5%Cu
d. 50% Al, 45%Zn and 5%Cu

144. The negative charge on clay surface and the swarm of positive counter ions are called as

a. Electric potential
b. Surface potential
c. Electric double layer
d. Zeta potential

145. Which one of the following has the - highest cation exchange capacity (weight basis)

a. Kaolinite
b. Illite
c. Humus
d. Montmorillonite

146. Most common method for determining soluble Na and Ca ion in soil is

a. Colorimetric
b. Flame photometric
c. Potentiometric
d. Titrimetric

147. The elements which ionize readily or form stable oxy-anions are

a. Lithophile elements
b. Chalcophile elements
c. Atmosphile elements
d. Biophile elements

148. The buffering capacity is higher in soils containing

a. 2:1:1 clay minerals
b. 2:1 clay minerals
c. 1:1 clay minerals
d. All of these

149. Aluminium in octahedral sheet of clay mineral indicates

a. Dioctahedral
b. Trioctahedral
c. Tetrahedral
d. Monohedral

150. Acid sulfate soils occur in the region of

a. Hilly areas
b. Alluvial plane areas
c. Forest area
d. Coastal areas

151. On weight basis which element occupies a major part (46 %) in earth's crust

a. O
b. Si
c. Al
d. Fe

152. Which type of mineral Chlorite is

a. 2:1:1 type
b. 2:1expanding type
c. 1:1 type
d. 2:1 non expanding type

153. Which one is the method for estimation of Gypsum Requirement

a. Olsen's method
b. Schoonover's method
c. Bray's method
d. Morgan's method

154. Which is important oxidizing agent used as standard solution

a. Oxalic acid
b. Potassium dichromate
c. Ferrous sulphate
d. Sodium thiosulphate

155. What common extractants are used to determine the availability of Fe, Mn, Zn, Cu in soil?

a. EDTA
b. DTPA
c. NH_4OAc
d. HCI

156. The permanent charge is more as compared pH-dependent charge in

a. Kaolinite
b. Humus
c. Montmorillonite
d. Hallosite

157. Formula for quick lime is

a. $CaCO_3$
b. $Ca(OH)_2$
c. CaO
d. $CaCO_3MgCO_3$

158. Choose the correct statement in given option

a. Illite clay particles are smaller than montmorillonite particles
b. Internal surface of kaolinite, clay is negatively charged
c. Cation exchange capacity of vermiculite is 2-15 Cmol/kg
d. Chlorites are silicates of magnesium

159. The SiO_2: Al_2O_3 ratio in lateritic soils varies between

a. 0-1
b. 2-4
c. 1-2
d. >4

160. In Kjeldahl digestion of soil for N estimation, K_2SO_4 is used to

a. Increase the temperature of digestion
b. Catalyses the reaction
c. Convert the NH_4^+ to NO_3^-
d. Break triple bond

161. In Kjeldahl digestion of soil for N estimation, which of the following reagent is used as catalyst

a. Na_2SO_4
b. K_2SO_4
c. $CuSO_4$
d. $CaSO_4$

162. One of the dimensions of nano particles must be within

a. 1 nm
b. 100 nm
c. 1000 nm
d. 100 mm

163. In Kjeldahl's method the organic N is digested with conc. H_2SO_4 and converted into

a. Organic-N
b. Nitrate sulphate
c. Ammonium sulphate
d. Sodium sulphate

164. Which group of colloid have highest cation exchange capacity by volume basis ?

a. Humus
b. Vermiculite
c. Smectite
d. Kaolinite

165. Which of the following phyllosilicate is dominant in bentonite?

a. Montmorillonite
b. Kaolinite
c. Dickite
d. Illite

166. The pH of AB-DTPA extractant is adjusted to

a. 7.0
b. 7.3
c. 4.0
d. 7.6

167. Which of the following mineral is not a polymorph?

a. Quartz
b. Tridymite
c. Crystobalite
d. Leucite

168. Which of the following minerals is likely to be chemically weathered most easily?

a. K-feldspar
b. Olivine
c. Quartz
d. Hornblende

169. Extraction of humic acid from soils is higher by the use of

a. Acid
b. Alkali
c. Acetone
d. Water

170. The amount of charge reaching to the central cation Na^+ from an anion in a cubic coordination is

a. -(1/8)
b. -(1/6)
c. +(1/8)
d. +(1/6)

171. Ligand adsorption of P by soil results in

a. Increase of CEC
b. Increase of AEC
c. Decrease of pH
d. Increase specific surface area

172. The polyphenol theory postulated by Flaig is related to which component in soil

a. Soil colloids
b. Soil enzymes
c. Soil air
d. Soil organic matter

173. Linear portion of Q/I curve of K represents

a. K adsorbed on planar, relatively high energy sites
b. K adsorbed on planar, relatively low energy sites
c. K fixed at geometrically selective adsorption sites
d. Non-exchangeable K

174. Solution A has pH of 3 and solution B has pH of 5. The active acidity of solution A is how many times greater than that of solution B

a. 2 times
b. 10 times
c. 20 times
d. 100 times

175. The EC of irrigation water is 1000 micro mhos/cm, which equals to

a. 100 ppm
b. 6400 ppm
c. 640 ppm
d. 2000 ppm

176. Lime is used as an amendment to reclaim :

a. Sodic soils
b. Acid soils
c. Calcareous soils
d. Saline soil

177. Which of the following soils has higher buffering capacity and lower activity ratio of potassium

a. Kaolinite dominant soil
b. Smectite dominant soil
c. Sandy soil
d. Red soil

178. In trioctahedral arrangement of phyllosilicate minerals, three octahedral positions are filled by which of the following cation

a. Si^{4+}
b. Al^{3+}
c. Mg^{2+}
d. Fe^{3+}

179. Which of the following methods for estimation of available phosphate in soils is suitable for a soil having pH 5.5 and CEC being in moderate range

a. Bray P1method
b. Olsen's method
c. Mehlic method
d. AB-DTPA method

180. The logarithm of the negative pressure head in centimeter height of water is

a. pF
b. pH
c. Eh
d. Gh

181. If 'k' is intrinsic permeability and 'f' is fluidity then hydraulic conductivity is given by

a. kf
b. k/f
c. $\sqrt{kf}$
d. $\sqrt{k/f}$

182. Oxidation reduction potential of typical soil systems is

a. Not affected by pH
b. Equal at pH 5 and 7
c. Higher at pH 5 than 7
d. Lower at pH 5 than 7

183. Which one of the following fertilizers has zero equivalent acidity

a. Ammonium sulphate
b. Ammonium nitrate
c. Anhydrous ammonia
d. Calcium ammonium nitrate

184. Hallow cathode lamp is used in which of the following instruments

a. ICAP-AES
b. Flame photometer
c. AAS
d. AES

185. Permissible limit of RSC (me L^{-1}) in irrigation water up to which gypsum application is not recommended

a. 2.5
b. 3.5
c. 4.5
d. 5.5

186. The 2:2 type of mineral is the characteristics of

a. Vermiculite
b. Chlorite
c. Montmorillonite
d. Illite

187. Which one is purely a chemical process

a. Ammonia volatilization
b. Denitrification
c. N- fixation
d. Nitrification

188. Gypsum is used as amendment for the reclamation of

a. Saline soil
b. Sodic soil
c. Calcareous soil
d. Acid sulphate soil

189. Alkaline soils have the characteristics of :

a. EC > 4 mmhos/cm, pH > 8.5 and ESP >15
b. EC < 4 mmhos/cm, pH > 8.5 and ESP >15
c. EC > 4 mmhos/cm, pH < 8.5 and ESP >15
d. EC < 4 mmhos/cm, pH < 8,5 and ESP <15

190. Methanogenesis in waterlogged soil takes place at Eh of

a. < -200 mV
b. 200-300 mV
c. 300-400 mV
d. > 400 mV

191. Which one of the following element ionic potential is maximum

a. Na
b. Cs
c. K
d. Li

192. Ligand adsorption of a weak acid on oxide surface will be maximum when

a. pH is greater than pK value of the acid
b. Acid anion and undissociated acid are present in equal amounts
c. Undissociated acid is 10% of dissociatedacid
d. pH is less than pK value of the acid

193. A soil which has pH <8.5, ESP <15 and EC>4 dS/m at 25 °C in saturated condition is called

a. Saline-alkali
b. Saline
c. Alkali
d. Normal

194. One molecule of $K_2Cr_2O_7$ when react with H_2SO_4 gives three atoms of oxygen, the equivalent weight of $K_2Cr_2O_7$ is

a. 1/3rd of molecular weight
b. 1/4th of molecular weight
c. 1/6th of molecular weight
d. 1/12th of molecular weight

195. The type of weathering needing water for chemical reaction is called

a. Hydration
b. Hydrolysis
c. Dissolution
d. Oxidation

196. Flame photometer is used in the determination of

a. Nitrogen
b. Phosphorus
c. Potassium
d. Boron

197. The cadmium toxicity in the soil-plant system can be reduced to some extent by the application of

a. Zinc
b. Phosphorus
c. Potassium
d. Sulphur

198. The extractant used for available phosphorus in soils is

a. Alkaline permangnate
b. 0.5 M $NaHCO_3$
c. 1N Ammonium acetate
d. 1 N Acetic acid

199. pH of Morgan's reagent is

a. 3.8
b. 4.8
c. 5.8
d. 6.8

200. The CHNS analyzer is used for the determination of

a. Organic carbon
b. Total carbon
c. In organic carbon
d. Organic matter

201. Micronutrient contents in soils are measured by which instrument?
   a. Atomic absorption spectrophotometer
   b. Flame photometer
   c. Colorimeter
   d. Conductivity meter

202. Diethylene triamine penta acetic acid (DTPA) is used for estimation of :
   a. Ca, Mg and S
   b. N, P and K
   c. Fe, Mn, Cu and Zn
   d. B, Mo and Cl

203. Vermiculite is similar to illite except vermiculite has
   a. A higher CEC
   b. A lower CEC
   c. Not an expansive layer
   d. K-bridge between the layers

204. If the pH of a solution is 7, what is the pOH
   a. 4
   b. 10
   c. 7
   d. 5

205. Karl Fischer reagent used in the chemical method of determination of moisture in fertilizers contains
   a. Pyridine
   b. Pyrimidine
   c. Phthalic acid
   d. Phenolphthalein

206. Basic slag is produced when
   a. Iron ore is converted to pig iron
   b. Pig iron is converted to steel
   c. Steel is converted to an alloy
   d. Any of the above

207. Isomorphous substitution of Al^{3+} for some Mg^{2+} ions in the dioctahedral sheet accounts for most of the negative charge in
   a. Illite
   b. Smectite
   c. Allophane
   d. Hemetite

208. The acid sulphate soils are also called as
   a. Cat clays
   b. Red clays
   c. Black clays
   d. Brown clays

209. Of the fallowing humic substances, which is insoluble in both acid and alkali:

a. Humic acid
b. Fulvic acids
c. Humin
d. All

210. When phenolphthalein is added to an acid solution the solution becomes

a. Yellow
b. Colourless
c. Blue
d. Pink

211. Laterization is the process of accumulation of

a. Sesquioxides
b. Silica
c. $CaCO_3$
d. Salts

212. When phenolphthalein is added to an basic solution the solution becomes

a. Yellow
b. Colourless
c. Blue
d. Pink

213. Which element imparts red colour to the soil

a. Ca
b. Mg
c. Fe
d. S

214. Hydrogen bonding results in the interlayer of

a. Muscovite
b. Montmorillonite
c. Pyrophyllite
d. Kaolinite

215. Strength of H_2SO_4 used in the laboratory is

a. 36 N
b. 16 N
c. 12 N
d. 14 N

216. Size of K^+ ion is similar to that of

a. Na^+
b. Zn^{2+}
c. NH_4^+
d. Mg^{2+}

217. Black colour of Vertisol is due to the presence of

a. Organic matter
b. Titanium
c. Selenium
d. Montmorillonite clay

218. The calcium carbonate equivalent of $CaCO_3$ is

a. 179
b. 108
c. 136
d. 100

219. The potential difference between the fixed part and freely mobile portion of the diffuse double layer is known as

a. Stream potential
b. Matric potential
c. Zeta potential
d. Ionic potential

220. Organic matter in soil can be decomposed by:

a. Hydrogen peroxide
b. Sodium carbonates
c. Sodium hydroxide
d. Sulphuric acid

221. Which of the following elements cannot be determined by a flame photometer?

a. Sodium
b. Potassium
c. Calcium
d. Zinc

222. "Lattice hole" theory of potassium fixation was proposed by

a. F.E. Bear
b. M.L. Jackson
c. Page and Baver
d. McLean

223. Gram equivalent of a solute dissolved in 1000 ml of solvent is

a. Molar solution
b. Molal solution
c. Normal solution
d. Standard solution

224. Humic acids are soluble in

a. Alkali
b. Acid
c. Alcohol
d. Water

225. Alkaline permanganate method is used for the estimation of

a. Total nitrogen
b. Available nitrogen
c. Ammonical nitrogen
d. Nitrate nitrogen

226. Atomic Absorption Spectrophotometer is used for which analyses

a. Hydrocarbon analyses
b. Phosphate analyses
c. Pesticides analyses
d. Heavy metal analyses

227. Hydrolysis of urea in saline soils yields

a. Amides
b. Ammonium carbamate
c. Uric acids
d. Hydroxylamine

228. What is the half-life of ^{32}P

a. 32 days
b. 86.7 days
c. 14.3 days
d. 245 days

229. Soil structure is more stable when the dominant clay mineral is

a. Montmorillonite
b. Illite
c. Chlorite
d. Kaolinite

230. For chloride determination in plant samples, titration with standard silver nitrate solution is carried out at pH

a. 5.6
b. 6.0
c. 7.0
d. 8.2

231. Devarda's alloy contains (50:45:5)

a. Cu, Mn and Zn
b. Cu, Al and Zn
c. Cu, Mn and Al
d. Cu, Mg and Al

232. Buffering capacity of soil means resistance to a change in

a. Soil aeration
b. Soil pH
c. Soil microbial activity
d. Water saturation

233. Which one is the correct form of Nernst equation

a. Eh = E -(0.059/n) log (oxidized molecule)/(reduced molecule)$(H^+)^m$
b. Eh = E +(0.059/n) log (oxidized molecule)/(reduced molecule)$(H^+)^m$
c. Eh = E -(0.059/n) log (oxidized molecule)/(reduced molecule)$(H^+)^m$
d. Eh = E -(0.059/n) log (reduced molecule)/(oxidized molecule)$(H^+)^m$

234. The inorganic and organic labile P fraction of soil are collectively called

a. Intensity factor
b. Quantity factor
c. Capacity factor
d. None of these

235. Soil calcium carbonate loses CO_2 to form calcium oxide at

a. 105 °C
b. 105-500 °C
c. 350 °C
d. ≥770 °C

236. Graphite furnace atomic absorption spectrophotometry uses maximum temperature up to

a. 500 °C
b. 1000 °C
c. 2000 °C
d. 3000 °C

237. Neutral normal ammonium acetate is used for extraction of

a. Nitrate
b. Chloride
c. Potassium
d. Sulphur

238. Target pH value of liming is usually

a. 5.5 - 6.5
b. 5.0 - 5.5
c. 4.5 - 5.5
d. 4.5 - 5.0

239. Which of the following species is dominant in solution at soil pH value of 4.0

a. Al^{3+}
b. $Al\ OH)^{2+}$
c. Al (OH)
d. $Al\ (OH)^{3+}$

240. Chemical formula of pyrite reclaiment is

a. FeS
b. CuS
c. MnS
d. FeS_2

241. Most important metal ions responsible for the binding of phosphates in soils are

a. Al, Fe and Ca
b. Al, Fe and K
c. Fe, Ca and Mg
d. Al, Fe and Zn

242. In which type of soil Olsen's method for phosphorus determination is used

a. Acidic soil
b. Saline soil
c. Neutral and calcareous soil
d. Degraded alkali soil

243. Major greenhouse gases based on their contribution to global warming are

a. $N_2O>CH_4>O_3>CO_2$
b. $CH_4>CO_2>O_3>N_2O$
c. $O_3>N_2O>CH_4>CO_2$
d. $CO_2>CH_4>O_3>N_2O$

244. The DTPA extractant concentration is used to determine micro-nutrients

a. 0.0005 M
b. 0.05 M
c. 0.005 M
d. 0.5 M

245. Identify the 1:1 clay mineral of soil among the following :

a. Vermiculite clay
b. Kaolinite clay
c. Montmorillonite clay
d. Chlorites clay

246. The general relative order of anion exchange is

a. $NO_3^- > OH^- > H_2PO_4^- > SO_4^{--}$
b. $OH^- > H_2PO_4^- > SO_4^{--} > NO_3^-$
c. $SO_4^{--} > NO_3^- > OH^- > H_2PO4^-$
d. $H_2PO_4^- > SO_4^{--} > NO_3^- > OH^-$

247. The anion exchange capacity (AEC) increases as soil pH

a. Increases
b. Decreases
c. First increases then decreases
d. Remain unaffected

248. In case of degraded alkali soil, there is a leaching of saline-sodic soil occurs and part of exchangeable Na is frequently replaced by

a. Ca^{++}
b. Mg^{++}
c. H^+
d. K^+

249. U.S. Salinity Laboratory developed the concept of SAR, that is equal to

a. $Na^+/\sqrt{(Ca^{++} + Mg^{++})/2}$
b. $Na^+ /(Ca^{++} + Mg^{++})/2$
c. $Mg^{++}/\sqrt{(Ca^{++} + Na^+)/2}$
d. $Ca^{++}/\sqrt{(Na^+ + Mg^{++})/2}$

250. The amount of iron sulphate ($FeSO_4 \cdot 7H_2O$) in tons is equivalent to 1.0 ton (one ton) of gypsum to improve soil

a. 1.0 ton
b. 1.62 ton
c. 0.57 ton
d. 0.80 ton

251. Two molecules that have the same chemical formula but are mirror images of each other are called

a. Enantiomers
b. Isomers
c. Isotopes
d. Mirrormers

252. The irrigation water contains CO_3^{--}, HCO_3^{-}, Ca^{++} and Mg^{++} ion 1, 4, 2.5 & 1.5. Calculate the RSC in irrigation water (mel/l).

a. 0.1
b. 10
c. 0.01
d. 1.0

253. Conversion factor of organic carbon to organic matter

a. 1.720
b. 1.750
c. 2.205
d. 1.724

254. The C-axis spacing of smectite group of silicate clays is

a. 0.66 nm
b. 0.72 nm
c. 1.01 nm
d. 1.40 nm

255. Walkley & Black method is used to determine

a. N_2
b. K_2O
c. OC
d. P_2O_5

256. What is the normality of concentrated hydrochloric acid (AR grade) in general

a. 40
b. 30
c. 11
d. 20

257. Available phosphorus (Olsen's method) and available potassium extracted from soil in laboratory respectively

a. Sodium acetate and $NaHCO_3$
b. Na_2CO_3 and Morgan reagent
c. $NaHCO_3$ and acetic acid
d. $NaHCO_3$ and Ammonium Acetate

258. By increasing the electrolytic concentration, the thickness of diffuse double layer

a. Increases
b. Decreases
c. Remains unaffected
d. Depends on pH

259. The alkaline potassium permanganate method (Alk. $KMnO_4$) for determination of available nitrogen (N) in soil was given by

a. Olsen
b. Bray
c. Subbiah and Asija
d. Subbiah and Makhija

260. The determination of micro-nutrients on atomic absorption spectrophotometer by DTPA was given by

a. Walkley and Black
b. Lindsay and Norwell
c. Bray's
d. Olsen's

261. The case of replacement of Ca^{2+} from colloids by plant uptake varies in order

a. Kaolinite > Illite > Montmorillonite > Peat
b. Montmorillonite > Peat > Kaolinite > Illite
c. Peat > Kaolinite > Illite > Montmorillonite
d. Peat > Illite > Kaolinite > Vermiculite

262. The molarity of concentrated H_2SO_4 (AR grade) is

a. 10 M
b. 12 M
c. 16 M
d. 18 M

263. Tannery is treated with chromium (Cr) which form of chromium is more hazardous to animal and human being

a. Cr^{2+}
b. Cr^{3+}
c. Cr^{6+}
d. None of these

264. Molybdate reagent is used for estimation of

a. N
b. P
c. Ca
d. B

265. Metal finishing industries effluents are rich in which heavy metals

a. Cr and Ni
b. Zn and Fe
c. Cu and Pb
d. Zn and Cu

266. Intensity of blue colour in phosphorus determination is measured at a wavelength of

a. 420nm
b. 470nm
c. 540nm
d. 660nm

267. The itai-itai disease is caused by higher content of which heavy metal in rice ?

a. Pb
b. Ni
c. Cd
d. Cr

268. Nuclear Power Plant accident took place which place in then USSR causing great hazard to agricultural soil and aquatic life

a. Beijing
b. Moscow
c. Chernobyl
d. Kicve

269. Which of the following gas (green house gas) contribute around 60% to the anthropogenic green house effect

a. N_2O
b. CO_2
c. SO_2
d. NH_3

270. Chlorite are the minerals belonging to

a. Tectosilicates
b. Cyclosilicates
c. Inoslicates
d. Phyllosilicates

271. Which one of the following soils is called White Alkali

a. Saline soils
b. Saline-sodic soils
c. Sodic soils
d. Acid sulphate soils

272. A soil contains minerals representing1:1 lattice clay. The mineral is

a. Smectite
b. Kaolinite
c. Illite
d. Vermiculite

273. Azomethine-H is used in determination of

a. Molybdenum
b. Sulphur
c. Boron
d. Calcium

274. Serpentine is a mineral of

a. Trioctahedral
b. Octachedral
c. Mixedlayer
d. Amorphous

275. 2:1:1 type of clay mineral contain an extra sheet of as compared to 2:1 type of clay minerals.

a. Gibbsite
b. Pyroxenes
c. Brucite
d. Biotite

276. Hexagonal cavity in clay minerals for the fixation of potassium and ammonium ions is surrounded by

a. Six oxygen atoms
b. Ten oxygen atoms
c. Eight oxygen atoms
d. Twelve oxygen atoms

277. The correct order of preference of anions for positive sites is

a. $Cl^{-} > PO_4^{3-} > SO_4^{2}$
b. $Cl^{-} >, SO_4^{2-} > PO_4^{3-}$
c. $PO_4^{3-} > SO_4^{2-} > Cl^{-}$
d. $SO_4^{2-} > PO_4^{3-} > Cl^{-}$

278. The name of the hydrous oxide clay having the formula Fe_3O_4 is

a. Hematite
b. Limonite
c. Magnetite
d. Geothite

279. Walkley and Black's method of organic matter determination allows .... per cent recovery

a. 54
b. 64
c. 74
d. 84

280. On flooding, most of the soils show a sharp drop in redox potential within a few

a. Weeks
b. Days
c. Months
d. Hours

281. Acid sulphate soils are formed mainly due to

a. Reduction of sulphates
b. Oxidation of sulphides
c. Evolution of hydrogen sulphide
d. Rice cultivation using ammonium sulphate

282. Which one of the following is responsible for the dispersion of soil particles

a. Exchangeable potassium
b. Exchangeable calcium
c. Exchangeable magnesium
d. Exchangeable sodium

283. Which silicate mineral has the lowest CEC among below

a. Montomorillonite
b. Kaolnite
c. Chlorite
d. Vermiculite

284. The clay mineral having highest CEC is

a. Illite
b. Vermiculite
c. Montmorillonite
d. Chlorite

285. The derivation of the laws governing equilibrium constants come from

a. Thermodynamics
b. Geo Science
c. Hydrolysis
d. Entropy

286. Eriochrome Black-T indicator used for determination of

a. Lime requirement
b. Gypsum requirement
c. Organic C
d. Boron determination

287. In1:1 type of silicate clay minerals, two units are joined together by

a. O-O linkage
b. O-OH linkage
c. O-K-O linkage
d. OH-OH strong linkage

288. Permanent negative charges on silicate clay minerals are developed due to

a. Broken edges
b. Phenolic groups
c. Isomorphous substitution
d. Dissociation of (OH) ions

289. What could be the ESP of a soil, which ha, CEC=25 cmol (P^{+})/kg and exch. Na=5 cmol (p^{+})/kg

a. 5.0
b. 20.0
c. 15.0
d. 25.0

290. A soil having pH>8.5, ESP>15 and EC<4 dS/m at 25°C is called

a. saline soil
b. Alkali soil
c. Non saline alkali
d. Normal soil

291. The radiation source used in atomic absorption spectrophotometry is

a. Hollow cathode lamp
b. Shallow cathode lamp
c. Hollow anode lamp
d. Dutarium lamp

292. Wide cracks commonly form in clay like

a. Kaolinite
b. Smectite
c. Saponite
d. Montmorillonite

293. Chemical formula of siderite is

a. $CaCO_3$
b. Fe_2O_3
c. $Ca3(PO_4)^3$
d. $FeCO_3$

294. For reclamation of acid soils, amendment used is

a. $CaSO_4$ $2H_2O$
b. $CaCO_3$
c. H_2SO_4
d. $CaCl_2$

295. Hydrogen bonding takes place in the interlayer of

a. Kaolinite
b. Montmorillonite
c. Mica
d. Chlorite

296.Which one is an amorphous mineral

a. Kaolinite
b. Illite
c. Allophane
d. Smectite

297.Which element is the only non metal among the micronutrients

a. Manganese
b. Boron
c. Iron
d. Zinc

298. Most of the clay minerals in soil belong to

a. Tectosilicates
b. Orthosilicates
c. Phyllosilicates
d. Inosilicates

299. Which index proved to be useful to assess the dominating type of clay mineral in soil proposed by Martin and Russell

a. Y- index
b. A-value
c. CYI
d. E -value

300. In order of decreasing resistance to weathering the correct group out of the following is

a. Muscovite-Biotite-Anorthite
b. Anorthite-Muscovite-Biotite
c. Biotite-Muscovite-Anorthite
d. Anorthite-Biotite- Muscovite

301. Calcium hydroxide has a neutralizing value (calcium carbonate equivalent) of

a. 179 %
b. 136 %
c. 109 %
d. 100 %

302. The pH value of soil 'A' and soil 'B' are 4.0 and 6.0, respectively, how many times the soil A has more H^+ than soil B

a. Soil A has 10 times more H^+ than soil B
b. Soil A has10 times less H^+ than soil B
c. Soil A has100 times more H^+ than soil B
d. Soil A has100 times less H^+ than soil B

303. Who is considered as father of soil testing

a. Colwell
b. Troug
c. Parker
d. Puri

304. The calcium carbonate equivalent to neutralize acidity of 100 kg of urea fertilizer is

a. 110
b. 55
c. 148
d. 84

305. Acidic conditions developed due to leaching of bases favour the formation of minerals.

a. Illite
b. Montmorillonitie
c. Kaolinite
d. Smectite

306. The high temperature, intense leaching and basic kind of parent material all favour the removal of

a. Sesquioxides
b. Clay
c. Silica
d. Metallic cations

307. The pH scale was proposed by

a. Sorenson
b. Munsell
c. Arnon
d. Liebig

308. Which of, the following mineral is non expanding 2 :1 type mineral

a. Montmorillonite
b. Nontronite
c. Beidellite
d. Illite

309. The pH meter works on the basis of

a. Laplas equation
b. Winstor equation
c. Bohr's equation
d. Nernst's equation

310. A quantitative insight into relationship of light absorption to the concentration is defined by

a. Beer-Lambert law
b. Goldschmidt's low
c. Vant Hoffs law
d. Ratio law

311. Cation exchange capacity (CEC) of mica is

a. 0
b. 8 - 30
c. 30- 50
d. 150-200

312. The suspension stability of mineral colloid as compared to organic colloid is generally

a. Higher
b. Lower
c. Nearly similar
d. Not specific

313. The net charge on humus varies, with soil pH. The negative charge on humus is extremely high in soils with pH of

a. 7-10
b. >10
c. 3-6
d. <3

314. The redox potential of a submerged rice soil varies between.... to..... millivolts

a. +500 to +700
b. +l to -100
c. +100 to +200
d. -300 to -400

315. A soil having 40 cmol (p+)/kg CEC holds 16, 6, 2 and 2 cmol (p+)/kg of Ca^{++}, Mg^{++}, K^{+} and Na^{+} respectively will have base saturation of

a. 62%
b. 65%
c. 60%
d. 58%

316. A soil having pH=8.0, EC=5.2 dS/m, CEC=30 cmol(p+)/kg and 5 cmol(p+)/kg Na will be designated as

a. Saline soil
b. Sodic soil
c. Saline-sodic soil
d. Degraded sodic soil

317. From Q/I relationship described by Beckett (1964), a high PBC^{K} indicates that

a. The soil has been recently limed
b. The soil needs potassium fertilization
c. Availability of potassium in the soil is excellent
d. The soil needs liming

318. In cation exchange,18 mg of NH_4 can be exchanged for (meq concept)

a. 6 mg of Al
b. 12 mg of Mg
c. 46 mg of Na
d. 40 mg of Ca

319.In organic-colloids of soil, the ion exchange is due to

a. -COOH group
b. -OH (phenolic group)
c. -COOH and -OH (phenolic group)
d. Isomorphous substitution

320. The RSC of water containing carbonate, bicarbonate, calcium, magnesium and sodium as 5, 35, 20, 10, and 10 meq/L, respectively will be

a. 0 meq/L
b. 20 meq/L
c. 10 meq/L
d. 25 meq/L

321. The C-axis spacing of chlorite mineral is

a. 7 A° b. 10 A°

c. 14 A° d. 18 A°

322. In terms of redox-potential of soils (mv) represents a highly reduced condition of the soil

a. -250 b. -100

c. + 200 d. + 600

323. A solution with pH=6.0 , the normality of H^+ would be

a. 0.00006 b. 1×10^{-7}

c. 1×10^{6} d. 0.000001

324. If saturation extract of a soil has EC=2.0 dS/m, what would be its osmotic potential in bars

a. -0.18 b. -0.36

c. -0.72 d. -1.08

325. In a soil, the CEC=20 meq/100g and exchangeable H^+=3 meq/100g, than the percent base saturation would be

a. 60% b. 15%

c. 85% d. 40%

326. Which extractant is most commonly used for estimation of available zinc in soils

a. 0.05 N HCl b. 1 NH_4OAC (pH 4.6)

c. 0.1N HCl d. 0.005 M DTPA+0.1 M TEA+0.01 M $CaCl_2$ (pH 7.3)

327. The dispersion phenomenon by the addition of sodium salt solution is related with

a. Soil water potential b. Osmotic potential

c. Zeta potential d. Chemical potential

328. If the contents of exchangeable ions (cmol/kg) determined in soil are $Na^+ = 5$, $K^+ = 5$, $Ca^{2+} = 10$, $Mg^{2+} = 2$, $H^+ = 1$, $Al^{3+} = 1$, $CO_3 = 1.6$ and $HCO_3 = 0.4$, the CEC of soil is

a. 22 cmol (P+) kg^{-1}
b. 24 cmol (P+) kg^{-1}
c. 26 cmol (P+) kg^{-1}
d. 12 cmol (P+) kg^{-1}

329. Single super phosphate is manufactured by treating rock phosphate with

a. Hydrochloric acid
b. Nitric acid
c. Sulphuric acid
d. Phosphoric acid

330. Aluminium ions contribute to soil acidity through their tendency to

a. Solubilise
b. Hydrolyze
c. Oxidize
d. Reduce

331. Which nutrient element is involved in eutrophication of lakes and surface water bodies

a. Zinc
b. Phosphorus
c. Potassium
d. Sulphur

332. The most common extracting agent used in estimation of inorganic nitrogen is

a. 1N NH_4OAC
b. 0.15% $CaCl_2$
c. O.5 M $NaHCO_3$
d. 2 M KCI

333. For irrigation with sewage water, its BOD should be within the range of

a. 100-150 ppm
b. 200-350 ppm
c. 1-10 ppm
d. 400-500 ppm

334. At high pH values the CEC of humus on mass basis is

a. 80-100 cmol/kg
b. 100-150 cmol/kg
c. 150-300 cmol/kg
d. 350-500 cmol/kg

335. Lime stone is the chief source for manufacturing of

a. Cement
b. Phosphatic fertilizer
c. Steel
d. Basic slag

336. The unit of expression of cation exchange capacity is

a. meq/100 mg
b. cmol (P^+)/g
c. dS/m
d. None of this above

337. The clay mineral having highest cation exchange capacity is

a. Kaolinite
b. Vermiculite
c. Illite
d. Chlorite

338. Ligand adsorption of weak acid anion is maximum at a pH which is

a. Equal to pK of weak acid
b. Greater than 10 times of pK
c. Less than 10 times of pK
d. Independent of pK

339. The source of permanent negative charge on silicate clays is

a. Isomorphous substitution
b. Hydrolysis of water
c. Ionization of OH^- groups
d. Alteration

340. The titrant used in the determination of calcium and magnesium is

a. DTPA
b. EDTA
c. NH_4OAc
d. $FeSO_4$

341. Base unsaturated soils are

a. Acid soils
b. Desert soils
c. Calcareous soils
d. Submerged soils

342. The dispersing agent used in soil particle size analysis is

a. Sodium hexametaphosphate
b. Calcium chloride
c. Hydrogen peroxide
d. Sulphuric acid

343. DTPA stands for

a. Diethylene Triamine Penta Aacetic acid
b. Dichlor Triamine Penta Acetic acid
c. Dichloro Triethylene Penta Acetic acid
d. Diethylene Trichloro Penta Acetic acid

344. Absorbance in a spectrophotometer is equal to

a. Po/P
b. 1/T
c. Log Po/P
d. Log P/Po

345.Aqua regia is prepared by mixing

a. Concentrated HCl and HNO_3
b. Concentrated HCl and H_2SO_4
c. Concentrated HNO_3 and H_2SO_4
d. Concentrated $HClO_4$ and H_2SO_4

346. Which one is not a radio isotope

a. 35S
b. 137Cs
c. 14C
d. 15N

347. Critical limit of available phosphorus in soil for wheat crop by Olsen's method Is

a. 5 kg/ha
b. 10 kg/ha
c. 15 kg/ha
d. 22.5 kg/ha

348.Diphenylamine is an indicator, which is used for determination of

a. Organic carbon
b. Micro nutrients
c. Calcium
d. Posphorus

349. The immobility of phosphorus in soils is primarily a result of orthophosphates

a. Reacting with Fe and Al in acid soils and Ca in, neutral and basic soils
b. Reacting with K in acid soils and Ca in neutral and basic soils
c. Reacting with Ca in acid soils and Fe and 'Al in neutral and basic soils
d. Being immobilised in organic matter

350. In acid soils phosphorus is fixed as

a. Fe and Al phosphate
b. Ca phosphate
c. Hydroxy apatite
d. Organic phosphate

351. Concentration of which element is highest in soil

a. Fe
b. O_2
c. Al
d. Si

352. NH_4 fixation takes place in

a. Vermiculite and illite
b. Montmorillonite and kaolinite
c. Chlorite and kaolinite
d. All of these

353. Which is a metalloid that is very toxic to organisms

a. Arsenic
b. Chromium
c. Mercury
d. Cadmium

354. Chemical formula of single superphosphate is

a. Ca (H_2PO_4)
b. $CaHPO_2$
c. $Ca^3(PO_4)$ (Cl) Ca (H_2PO_4)
d. $CaSO_4.2H_2O$

355. Phosphatic fertilizers are known to contain contaminants like

a. Copper and Chromium
b. Iron and Aluminium
c. Arsenic and Mercury
d. Cadmium and Fluoride

356. As per Beer-Lambert's law, the relationship between optical density (OD) and transmittance (%T) of a coloured solution is given by

a. OD = log T
b. OD = log 2.0-T
c. OD = 2.0 -log T
d. OD = log T-2.0

357. The concentration of HPO_4^{2-} and $H_4PO_4^-$ ions in solution is equal at a pH of

a. 5.6
b. 6.5
c. 7.2
d. 8.2

358. At what pH, there are approximately equal amounts of $H_2PO_4^-$ and HPO_3^{2-} ?

a. 6.5
b. 7.0
c. 7.2
d. 8.5

359. In the formation of organometallic complexes in soils the metal ion acts as a/an.

a. Ligand
b. Electron donor
c. Electron acceptor
d. Inert ion

360. Which one of the following formula is used for calculation of base saturation ?

a. Total bases/CEC x 100

b. Exchangeable bases/CEC x 100

c. CEC/total bases x 100

d. CEC/exchangeable bases x 100

361. Which position at the layer silicate has the maximum specificity for K^+ ion

a. a-position

b. p-position

c. i-position

d. Both a-and p-positions

362. In total N determination of a soil sample by Kjeldahl method, salicylic acid is added before digestion to include

a. NH_4-N

b. NO_3-N

c. Amide-N

d. Protein-N

363. Potassium activity ratio (AR^K_e) is expressed as

a. aK/√(aCa+aMg)

b. aK/√(aCa+aMg)/2

c. aK/(aCa+aMg)

d. aK/(aCa+aMg)/2

364. The DTPA-TEA-$CaCl_2$ extractant is designed to prevent excessive dissolution of

a. $CaSO_4$

b. CaO

c. $MgCl_2$

d. $CaCO_3$

365. Which extract is used for determination of CEC ?

a. 1N NH_4OAC

b. 1N $NaOHCO_3$

c. 1N KNH_3

d. Olsen reagent

366. Availability of boron high in soil basis of pH reaction

a. Alkaline soil

b. Saline soil

c. Acid soil

d. Neutral soil

367. The process of bringing bulk soil sample to the desired quantity is called

a. Reduction
b. Minimization
c. Quartering
d. Partitioning

368. For most minerals, the strength of cation adsorption, or lyotropic series, is

a. $Ca^{2+} > Al^{3+} > Mg^{2+} > K^{+} = Na^{+} > NH_4$
b. $Al^{3+} > Ca^{2+} > Mg^{2+} > K^{+} = NH_4 > Na^{+}$
c. $Mg^{2+} > Al^{3+} > Ca^{2+} > K^{+} = Na^{+} > NH_4$
d. $Al^{3+} > Mg^{2+} > Ca^{2+} > K^{+} = NH_4 > Na^{+}$

369. CEC of which clay mineral is highest (120-150 cmol/kg) ?

a. Mica
b. Vermiculite
c. Montmorillonite
d. Chlorite

370. The negative charge on 2 : 1 clays is

a. pH dependent
b. pH independent
c. Both (a) and (b)
d. None of these

371. 29.5 mg of an organic compound containing nitrogen was digested according to Kjeldahl's method and the evolved ammonia was absorbed in 20ml of 0.1 M HCl solution. The excess of the acid required 15ml of 0.1 M NaOH solution for complete neutralization. The percentage of nitrogen in the compound is

a. 23.7
b. 29.5
c. 47.4
d. 59.0

372. If 2 ml of 1N $K_2Cr_2O_7$, solution was consumed for oxidation of organic C present in 1 gram of a soil, the organic matter percentage of the soil is

a. 0.3
b. 0.6
c. 1.0
d. 1.5

373. Anion exchange capacity is highest in

a. Vermiculite
b. Montmorillonite
c. Kaolinite
d. Illite

374. Carbon-14 (^{14}C) is used for C dating. What is the half life of ^{14}C

a. 8370 years
b. 5730 years
c. 1620 years
d. 13 years

375. A soil sample indicating pH = 7.5, EC = 4.8 dS/m, exchangeable sodium' percentage = 12, may be categorised as which one of the following ?

a. Alkaline soil
b. Saline soil
c. Saline - alkaline soil
d. Normal fertile soil

376. Which of the minerals are Al-containing

a. Gibbsite
b. Albite
c. Boehmite
d. All of these

377. Choose the correct statement in given option

a. In soils both H_2PO_4 and HPO_4 ions exist at pH 3.0
b. Plants absorb phosphorus largely in the form of metaphosphate ion.
c. The concentration of various phosphate ions in soil solution is intimately related to pH.
d. In alkaline soils the product of phosphate fixation are largely complex phosphates of iron and aluminium

378. Choose the correct statement in given option

a. A single cation on a given clay mineral can only be held with a fixed bonding energy.
b. A given clay has fixed bonding energy for different cations.
c. Different clays have widely different bonding energies for a given cation.
d. The complimentary ion has no effects on bonding energies.

379. Which of the following are magnesium source also

1. Kainite
2. Sylvinite
3. Langbeinite
4. Carnallite

Select the correct answer using the codes:

a. (1), (3) and (4)
b. (2), (3) and (4)
c. (1) and (4)
d. (2) and (3)

380. Match List I with List II

a. Strengite	1. FeOOH
b. Variscite	2. $Al_2O_3.3H_2O$
c. Gibbsite	3. $FePO_4.2H_2O$
d. Hematite	4. $AlPO_4.2H_2O$

381. Match List I (Mineral) with List II (Chemical formula)

a. Geothite	1. $CaCO_3$. $MgCO_3$
b. Gypsum	2. Al_2O_3. $3H_20$
c. Gibbsite	3. Fe_2O_3
d. Dolomite	4. $CaSO_4.2H_2O$

382. Match List I (Available nutrient) with List II (Extractants used)

a. Molybdenum	1. Neutral normal ammonium acetate
b. Iron	2. Hot Water
c. Boron	3. 0.005 M DTPA (pH-7.3)
d. Potassium	4. Ammonium oxalate (pH- 3.3)

383. Match List I (Rocks) with List II (Elements)

a. Wulfenite	1. Mn
b. Smithsonite	2. Cu
c. Malachite	3. Zn
d. Pyrolusite	4. Mo

384. Match List I with List II

a. Karl Fisher reagent	1. P determination
b. EBT indicator	2. Available N estimation
c. Ascorbic acid	3. Available Mo estimation
d. Alkaline permanganate	4. Moisture determination in urea
e. Grigg's reagent	5. Ca estimation

385.Match List I with List II

List I	List II
a. Production of H_2 feed stock	1. Solvay's process
b. Single super phosphate	2. Froth floatation
c. Ammonium chloride	3. Haber-Bosch
d. Beneficiation of rock phosphate	4. Steam reforming
e. Ammonia synthesis	5. EID (Parry) Ltd.

386. Match List I with List II

List I	List II
a. Hornblende	1. SiO_2
b. Olivine	2. SiO_3
c. Feldspar	3. SiO_4
d. Pyroxene	4. Si_2O_5
e. Mica	5. Si_2O_3

387. Match List I with List II

List I	List II
a. Truog's reagent	1. pH7.3
b. Egner's reagent	2. pH 8.5
c. Morgan's reagent	3. pH4.8
d. Olsen's reagent	4. pH3.0
e. DTPA reagent	5. pH 3.8

388. Match List I with List II

List I	List II
a. Potentiometry	1. Sodium
b. Gas Chromatograph	2. Phosphorus
c. Flame Photometry	3. Volatile compound
d. Conductometry	4. pH
e. Colorimeter	5. EC

389.Match List I with List II

List I	List II
a. Ratio law	1. Salt-tolerance of crops
b. Bragg's law	2. Selectivity coefficient
c. ZPC	3. Donnan membrane equilibrium

d. Osmo-regulation — 4. X-ray diffraction

e. Gapon equation — 5. Variable charge surface

390. Match List I with List II

a. Inert elements	1. Zn
b. Amphoteric metal	2. Li
c. Lightest metal	3. Ozone
d. Light blue gas	4. Zero group

391.Match List I (Trade name) with List II (Chemical name)

a. Black ash	1. Zinc oxide
b. Chili saltpeter	2. Crude form of sodium carbonate
c. Chinese white	3. Copper sulphate
d. Blue vitriol	4. Sodium nitrate

392. Match List I with List II

a. Indicator used for estimation of Ca^2 + by EDTA method	1. $CaCO_3$
b. Indicator used for estimation of CO_3^{-2}	2. Sodium hexameta phosphate
c. Binding agent in soils	3. Mureoxide
d. Soil dispersing agent	4. Phenolphthalein

393. Match List I with List II

a. Precipitation indicator	1. Starch
b. Redox indicator	2. Mureoxide
c. Self indicator	3. $KMnO_4$
d. Metal ion indicator	4. Orthophenanthroline
e. Adsorption indicator	5. Ferric alum

394. Match List I with List II

a. Mehlich No. 1	1. 0.5 M $NaHCO_3$ (pH-8.5)
b. Bray P1	2. 2 M NH_4HCO_3+0.005 M DTPA (pH-7.6)

c. Morgan reagent	3. 0.5 N HCl+0.025 M H_2SO_4
d. Olsen reagent	4. 0.73 M NaOAc+7.4 M HAOc (pH-4.8)
e. AB-DTPA	5. 0.03 N NH_4F+0.025 M HCl

ANSWERS (Soil Chemistry)									
1.c	2.c	3.b	4.a	5.a	6.b	7.a	8.c	9.a	10.a
11.d	12.a	13.d	14.b	15.c	16.a	17.c	18.a	19.d	20.b
21.b	22.c	23.c	24.a	25.c	26.a	27.b	28.d	29.a	30.c
31.c	32.c	33.c	34.b	35.a	36.a	37.a	38.c	39.b	40.a
41.b	42.d	43.c	44.d	45.b	46.a	47.a	48.c	49.c	50.b
51.c	52.b	53.a	54.a	55.c	56.a	57.b	58.d	59.c	60.a
61.b	62.b	63.c	64.c	65.a	66.d	67.b	68.a	69.a	70.b
71.b	72.d	73.a	74.b	75.c	76.d	77.b	78.b	79.a	80.d
81.d	82.a	83.c	84.b	85.d	86.a	87.c	88.b	89.a	90.c
91.a	92.a	93.a	94.a	95.a	96.c	97.a	98.b	99.a	100. d
101.a	102.a	103.a	104.c	105.c	106.a	107.a	108.a	109.d	110.c
111.a	112.d	113.a	114.c	115.c	116.b	117.c	118.b	119.d	120.d
121.c	122.c	123.b	124.a	125.b	126.a	127.b	128.c	129.c	130.d
131.d	132.a	133.c	134.c	135.a	136.a	137.a	138.d	139.b	140.d
141.a	142.b	143.a	144.c	145.c	146.b	147.a	148.b	149.a	150.d
151.a	152.a	153.b	154.b	155.b	156.c	157.c	158.d	159.a	160.a
161.c	162.b	163.c	164.b	165.a	166.d	167.d	168.b	169.b	170.d
171.a	172.d	173.d	174.d	175.c	176.b	177.b	178.c	179.a	180.a
181.a	182.d	183.d	184.c	185.a	186.b	187.a	188.b	189.c	190.a
191.d	192.b	193.b	194.c	195.a	196.c	197.a	198.b	199.b	200.b
201.a	202.c	203.a	204.c	205.a	206.a	207.b	208.a	209.c	210.b
211.a	212.d	213.c	214.d	215.a	216.c	217.b	218.d	219.c	220.a
221.d	222.c	223.c	224.a	225.b	226.d	227.b	228.c	229.a	230.d
231.b	232.b	233.d	234.b	235.d	236.d	237.c	238.a	239.a	240.d
241.a	242.c	243.d	244.c	245.b	246.d	247.b	248.a	249.a	250.b
251.b	252.d	253.d	254.d	255.c	256.c	257.d	258.b	259.c	260.b
261.b	262.d	263.c	264.b	265.a	266.d	267.c	268.c	269.b	270.d
271.a	272.b	273.c	274.a	275.c	276.d	277.c	278.c	279.c	280a
281.b	282.d	283.b	284.b	285.a	286.b	287.b	288.c	289.b	290.b
291.a	292.d	293.d	294.b	295.a	296.c	297.b	298.c	299.a	300.a
301.b	302.c	303.b	304.d	305.c	306.c	307.a	308.d	309.d	310.a
311.a	312.b	313.c	314.d	315.b	316.a	317.c	318.d	319.c	320.c
321.c	322.a	323.d	324.c	325.c	326.d	327.c	328.b	329.c	330.b
331.b	332.d	333.a	334.c	335.a	336.d	337.b	338.a	339.a	340.b
341.a	342.a	343.a	344.c	345.a	346.a	347.b	348.a	349.a	350.a
351.b	352.a	353.a	354.a	355.d	356.c	357.b	358.c	359.a	360.a
361.c	362.b	363.a	364.d	365.a	366.a	367.c	368.b	369.b	370.c
371.b	372.b	373.c	374.b	375.b	376.d	377.c	378.c	379.a	
380. a.3 b.4 c.2 d.1					381. a.3 b.4 c.2 d.1				
382. a.4 b.3 c.2 d.1					383. a.4 b.3 c.2 d.1				
384. a.4 b.5 c.1 d.2 e.3					385. a.4 b.5 c.1 d.2 e.3				
386. a.2 b.3 c.1 d.5 e.4					387. a.4 b.5 c.3 d.2 e.1				

388. a.4 b.3 c.1 d.5 e.2	389. a.2 b.4 c.3 d.1 e.5
390. a.4 b.1 c.2 d.3	391. a.2 b.4 c.1 d.3
392. a.3 b.4 c.1 d.2	393. a.4 b.5 c.3 d.2 e.1
394. a.3 b.5 c.4 d.1 e.2	

5
Soil Physics

1. The smallest size soil separate is

   a. Sand
   b. Silt
   c. Clay
   d. Loam

2. In Stocks law 'η' denotes :

   a. Density
   b. Gravity
   c. Viscosity
   d. Velocity

3. The average particle density of mineral soil is

   a. 2.65 g/cc
   b. 1.40 g/cc
   c. 2.65 kg/cc
   d. 1.33 kg/cc

4. The best soil structure is

   a. Crumby
   b. Platy
   c. Prismatic
   d. Columiner

5. Choose the false statement

   a. Total water of soil is known as Holard
   b. Amount of soil water non-available to plant is known as Echard
   c. Amount of soil water available to plant is known as Chresard
   d. Amount of soil water available to plant hold between 15 to 30 bar

6. If wire is formable but soil mass is easily deformable, then the soil is

   a. Non-plastic
   b. Plastic
   c. Slightly plastic
   d. Very plastic

7. Total porosity of clay soil is

   a. Less than sandy soil
   b. More than sandy soil
   c. More or less same
   d. All above

8. Hydraulic conductivity of soil is governed by

a. Darcy's law
b. Ficks law
c. Boyl's law
d. Charle's law

9. The optimum moisture content for tillage is

a. 20 % of field capacity
b. 40 % of field capacity
c. 60 % of field capacity
d. 80 % of field capacity

10. Capillary water is held between the tensions of

a. 15-31 bar
b. 1/3- 31 bar
c. 1/3- 15 bar
d. 31-100 bar

11. Water in soil always move from its

a. Higher potential to lower potential
b. Lower potential to higher potential
c. Both way
d. All above are correct

12. The hydraulic conductivity in soil maximum at ________ condition

a. Saturated
b. Field capacity
c. Wilting point
d. Air dry

13. Which of the following does not affect soil erosion

a. Climate
b. Vegetation
c. Solar
d. Topography

14. Which of the following represents the purity or strength of colour in Munsell colour chart

a. Hue
b. Value
c. Chroma
d. Hue and Value both

15. Mineral particles having diameter of 2-0.2 mm are known as

a. Gravel
b. Coarse sand
c. Fine sand
d. Silt

16. Forces responsible for retention of soil water are

 a. Cohesion
 b. Adhesion
 c. Cohesion and Adhesion
 d. None of these

17. What is the composition of soil air

 a. O_2 - 20%, N_2-80% and CO_2-0.21%
 b. O_2 - 20%, N_2-77% and CO_2 -0.50%
 c. O_2 - 22%, N_2-76% and CO_2 -1.5%
 d. O_2 - 24%, N_2 -74% and CO_2 -2.0%

18. Choose the incorrect statement

 a. Clay particle are electronegative charged particle.
 b. Crumby soil structure has more bulk density than platy structure
 c. Particle density did not affected by pore space
 d. Fine texture soil has low bulk density than sand soil

19. The sprinkler system of irrigation is most suitable for

 a. Undulated soils
 b. Saline soils
 c. Clay soils
 d. Acidic soils

20. Which of the soil has higher buffering capacity

 a. Sandy soils
 b. Clay soils
 c. Loamy sand soils
 d. Loamy soils

21. The heat capacity of given material is equal to its specific heat

 a. Divided by the mass of given material
 b. Added by the mass of given material
 c. Multiplied by the mass of given material
 d. Subtracted by the mass of given material

22. Choose the correct statement

 a. The entry of water into the soil is percolation
 b. Vertical movement of water is known as infiltration

c. Permafrost it is layer where soil temperature is always > 10°C

d. The downward movement of water through the soil is called percolation

23. Most economic method of irrigation for orchard under water scarcity conditions is

a. Drip system
b. Sprinkler system
c. Flood system
d. Ring system

24. The diameter of silt particles in USDA system is

a. 0.02-0.002 mm
b. 0.05-0.002 mm
c. 0.01-0.001 mm
d. 0.05-0.005 mm

25. The arrangement of primary and secondary soil particles is called as :

a. Soil structure
b. Soil texture
c. Soil consistency
d. Soil plasticity

26. The average arable surface soil may be considered to have a particle density of

a. 1.63 Mg/m^3
b. 2.66 Mg/m^3
c. 1.33 Mg/m^3
d. 2.33 Mg/m^3

27. Neutron moisture probe estimates moisture in soil by detecting molecules of

a. Water
b. Oxygen
c. Hydrogen
d. Carbon

28. The sizes of silt fractions in soils in ISSS system are

a. 0.2-0.02 mm
b. 0.2-0.002 mm
c. 0.5-0.002 mm
d. 0.05-0.002 mm

29. Choose the correct statement

a. The viscosity of water at 20 °C is 10 Centipoise
b. Soil water potentials is always positive
c. Piezometer cannot used to measure the submergence potential
d. Soil and leaf water potentials are determined by psychrometer

30.Choose the false statement in given option

a. The maximum field capacity in clay soil

b. Plasticity is non-dynamic properties of soil

c. Density is physical properties of soil.

d. The maximum available water in silt loam soil

31. What will be the texture of soil if it contains 60% sand, 30% silt and 10% clay?

a. Silt loam | b. Sandy loam

c. Clay loam | d. Sandy clay

32. The microrelief of soils produced by expansion and contraction with changes in moisture is known as

a. Fluvial deposits | b. Gilgai

c. Mottling | d. Heaving

33. Stokes' law suggests the rate of fall of particle is directly proportional to the square of

a. Density of the particle | b. Density of the fluid

c. Viscosity of medium | d. Radius of the particle

34. Void ratio of soil is expressed a

a. The ratio of volume of pores to the soils

b. The ratio of volume of solids to the pore volume

c. The ratio of pore volume to the total volume

d. The ratio of total volume to the pore volume

35. Very fine angular blocky structure belongs to size

a. <5 mm | b. 10-12mm

c. 5-10 mm | d. 20-50 mm

36. Adhesion means

a. Attraction between similar particles

b. Attraction between soil particle

c. Attraction between dissimilar particles

d. Attraction between soil separates

37. Flocculation occurs with
    a. Increase in negative charges
    b. Decrease in positive charges
    c. Decrease in zeta potential
    d. Increase in zeta potential
38. Which type of soil structure faces severe problem of soil crusting
    a. Loamy sand
    b. Sandy loam
    c. Silty loam
    d. Silty clay loam
39. The relatively inert sand and silt fractions can be called the soil
    a. Skeleton
    b. Chalka
    c. Pokkali
    d. Piping
40. Diversion terraces are not suitable for slopes more than
    a. 12%
    b. 5%
    c. 1%
    d. 7%
41. The sunflower is indicator plant of
    a. Copper & PWP
    b. Boron & PWP
    c. Molybdenum & PWP
    d. Nitrogen & PWP
42. Which textural class of soil considered to be optimal for-plant growth
    a. Clay
    b. Loam
    c. Clay loam
    d. Silt Clay
43. By which method soil textural class is perfectly determined
    a. Field method
    b. Use of textural triangle
    c. Hydrometer method
    d. International pipette method
44. Mean weight diameter index is used for characterization of
    a. Soil structure
    b. Soil organic matter
    c. Soil texture
    d. Soil particles
45. The latent heat of vaporization of water at boiling point is
    a. 537 cal/gm
    b. 580 cal/gm
    c. 590 cal/gm
    d. 100 cal/gm

46. Rise in capillary water is because of

    a. Gravitational force
    b. Adhesive farce
    c. Cohesive force
    d. Adhesion, Cohesion and Surface tension

47. When soil water is at hydrostatic pressure greater than atmospheric pressure, its potential is called.

    a. Matric potential
    b. Submergence potential
    c. Gravitational potential
    d. Osmatic potential

48. The presence of solute in soil water

    a. Increases the soil water potential
    b. Decrease the soil water potential
    c. Do not affect the soil water potential
    d. None of these above

49. The first stage of water erosion is

    a. Splash erosion
    b. Gully erosion
    c. Rill erosion
    d. Sheet erosion

50. The formula $q=k.\Delta H/L$ is:

    a. Darcy's law
    b. Flux density
    c. Hydraulic gradient
    d. Fick's law

51. Hysteresis occurs because of

    a. Non uniformity in pore space
    b. Contact angle effect
    c. Ink-bottle effect
    d. All of these above

52. Gaseous exchange can take place in soil by

    a. Total gas pressure gradient
    b. Diffusion and Mass flow
    c. Diffusion
    d. Mass flow

53. In saturated soil, the value of metric potential is :

    a. Unity
    b. Zero
    c. Positive
    d. Negative

54. Who is the father of soil physics

a. Hillel
b. Richard
c. Dokuchaev
d. Thomas way

55. Osmotic presser equal to

a. $0.36 \times EC$ dS/m
b. $36 \times EC$ dS/m
c. $360 \times EC$ dS/m
d. $0.036 \times EC$ dS/m

56. Aggregates stability index is the difference between percent of

a. Clay and Sand
b. Silt and Sand
c. Clay and Silt
d. Clay and Moisture retain of soil

57. The pF scale was given by:

a. Sorenson
b. Atterberg
c. White and Beckett
d. Schofield

58. Which one is correct?

a. 1 bar $=10^6$ dynes/cm^2
b. 1 Pascal = 100 N/m^2
c. 1 Pascal = 10 bar
d. 1 Mega pascal = 100 bars

59. Which one of the following is not a physical factor affecting the type of farming-

a. Topography
b. Climate
c. Labour supply
d. Soil

60. Movement of water molecular from a solution with their higher concentration towards their lower concentration solution through a semi-permeable membrane is known as

a. Diffusion
b. Osmosis
c. Imbibition
d. Transpiration

61. Which one of the following is responsible for the dispersion of soil particles

a. Exchangeable calcium
b. Exchangeable magnesium
c. Exchangeable sodium
d. Exchangeable potassium

62. The diameter of clay particle is

a. <200 μ b. <20 μ

c. <2 μ d. <5 μ

63. The hydraulic conductivity of soil decreases in the following order

a. Ca^{2+} soil < K^+ soil < Na^+ soil b. Ca^{2+} soil > K^+ soil > Na^+ soil

c. Na^+ soil < K^+ soil < Ca^{2+} soil d. Na^+ soil > K^+ soil > Ca^{2+} soil

64. Soil particle size of silt soil ranges between

a. 2.0 mm to 0.2 mm b. 0.2 mm to 0.02 mm

c. 0.02 mm to 0.002 mm d. 0.002 mm to 0.0002 mm

65. The infiltration capacity of a soil is measured by a device called

a. Tensiometer b. Penetrometer

c. TDR d. Double ring infiltrometer

66. The water content on mass basis can be converted into volume basis by multiplying it with

a. Particle density b. Hydraulic conductivity

c. Bulk density d. Flux density

67. The silt particles transported by the action of wind are termed as

a. Aeolian b. Loess

c. Alluvium d. Colluvium

68. The movement of heat in soil is expressed as...... law

a. Darcy's law b. Fick's law

c. Stock's law d. Fourier's law

69. Ravines is the a broad form of which types of soil erosion

a. Sheet erosion b. Plate erosion

c. Gully erosion d. Big rill erosion

70. A clay soil must contain at least .... clay separate, to be designated a clay

a. 70 % b. 80 %

c. 40 % d. 35 %

71. Wind blown materials, comprised primarily of silt with some fine sand and clay, are called

a. Eolian b. Loess

c. Moraine d. Alluvial

72. How many agroclimatic zones in India

a. 10 b. 15

c. 20 d. 25

73. The attraction of water molecules for solid surface is called

a. Cohesion b. Adhesion

c. Binding d. Suction

74. Undulating topography of land surface have the slopes of

a. <2% b. 2-8%

c. 8-16% d. >16%

75. Gas exchange between soil and atmosphere occur by

a. Diffusion b. Mass flow

c. Both (a) & (b) d. Sun light

76. The factor responsible for formation of acid soil is

a. Low rainfall b. High rainfall

c. High evaporation d. Low rainfall and high evaporation

77. Energy source for passive active remote sensing is

a. Solar radiation b. Radar

c. Sensor d. Detector

78. 'K' in the universal loss equation represent

a. Slope factor b. Soil erodibility factor

c. Crop management factor d. Rainfall erosivity factor

79. For mechanical analysis of a soil, organic matter is destroyed by treating the soil with

a. H_2O_2 (6%) b. NaOH (1N)

c. $K_2Cr_2O_7$ (1M) d. HCl (1N)

80. Contour cultivation is done for

a. Nutrients conservation
b. Soil and water conservation
c. Insect control
d. None of these

81. 1 bar is equal to

a. 0.9869 atm
b. Weight of a 1020 cm water column
c. 75.01 cm high mercury column
d. All of these

82. Bulk and particle density of soil is expressed as

a. mg/kg
b. kg/m^2
c. Mg/m^3
d. kg/m

83. The surface tension of water at 20 °C and 25 °C is

a. 80 and 72.7 dynes/cm
b. 72.7 and 80 dynes/cm
c. 20 and 25 dynes/cm
d. 0 and 4 dynes/cm

84. Downward entry of water into the soil is called

a. Percolation
b. Infiltration
c. Permeability
d. Hydraulic conductivity

85. Soil Moisture is measured by

a. Tensiometer
b. TDR
c. Peizometer
d. Both a and b

86. The rise of water in a capillary tube is in what way related to the diameter/radius of the tube

a. Inversely
b. Directly
c. Both
d. None of these

87. 'K' in soil loss equation A = RKLSCP stands for

a. Soil Productivity factor
b. Soil leaching factor
c. Crop management factor
d. Soil erodibility factor

88. The rise of water in a capillary tube is .... proportional to surface tension of water

a. Inversely
b. Directly
c. Both
d. None of these

89. The blocky types soil structure generally found in the

a. A horizon
b. C horizon
c. B horizon
d. O horizon

90. The entry of water in to the soil is called

a. Infiltration
b. Percolation
c. Ground water
d. None of these

91. The downward movement of water through the soil is called

a. Infiltration
b. Percolation
c. Ground water
d. None of these

92. Diffusion of gases in between soil air and atmospheric air takes place following

a. Boyle's law
b. Charle's law
c. Fick's law
d. Fourier's law

93. Which type of structure is dominant in grasslands ?

a. Spheroidal
b. Platy
c. Columnar
d. Blocky

94. A flow hydrograph is a plot of

a. Stream discharge against time
b. Infiltration against time
c. Stream velocity against time
d. Distance against time

95. The ideal soil for crop cultivation

a. Loam
b. Clay loam
c. Sandy loam
d. Clay

96. At a particular soil moisture tension, which soils have the highest moisture content?

a. Sandy
b. Loamy
c. Silty loam
d. Clayey

97. The efficiency of field irrigation would be more in

a. Sandy loam
b. Loam
c. Clay loam
d. Silty loam

98. Vertical movement of water in soil is known as

a. Percolation b. Filtration

c. Infiltration d. Diffusion

99. In rainy season rain water may contain about ______ as CO_2

a. 0.25 % b. 0.45 %

c. 0.03 % d. 0.04 %

100. Shelter belts are used to control

a. Water erosion b. Wind erosion

c. Water logging d. Soil crushing

101. Cause of water stagnation of surface of alkali soil is

a. Deflocculation b. Aggregation

c. Floculation d. None of these

102. Which one of the following water potential in soil is always positive ?

a. Osmotic potential b. Mark potential

c. Gravitational potential d. None of these

103. The term water potential was introduced by

a. C. W. Rose b. R. K. Sc

c. A. Richards d. E. Buckingham

104. The bench terraces are constructed on the slopes more than

a. 12% b. 15%

c. 10% d. 8%

105. In Darc's law, the hydraulic gradient has the physical dimensions of

a. $L^{-1}L^{2}$ b. $M^{-1}L^{3}T$

c. $L^{3}T^{-1}$ d. LL^{-1}

106. Watersheds are classified into different categories based on area that the mili-watershed contains area in hectare

a. 50,000-2,00,000 b. 10,000-50,000

c. 1,000-10,000 d. 100-1,000

107. Which soil moisture range is optimum for tillage operation

a. Hard consistency
b. Friable consistency
c. Plastic consistency
d. Sticky consistency

108. The hydraulic conductivity and differential water capacity of a soil are 10^{-3} cm/sec and 10^{-5} /cm respectively. The soil water diffusivity is

a. 10^{-2} cm^2/sec
b. 10^{-8} cm^2/sec
c. 10^2 cm^2/sec
d. 10 cm^2/sec

109. The force exerted by the freezing of water is equivalent to about is:

a. 1465 tonnes/m^{-2}
b. 150 tonnes/ft^{-2}
c. Both (a) and (b)
d. None of these

110. Very poor, infiltration due to adverse soil physical properties are found in

a. Acid soil
b. Saline soil
c. Alkali soil
d. Saline alkali soil

111. The detection ability of Neutron Moisture Meter which captures slow moving neutrons contains

a. Boron
b. Americium
c. Helium
d. Radium

112. The height of rise of water in a capillary tube is directly proportional to the

a. Surface tension of water
b. Radius of the tube
c. Difference of density of water
d. Acceleration of water

113. The......saturated montmorillonite has the highest plastic limit as compared to the other cation saturated montmorillonite clay.

a. Na
b. K
c. Ca
d. Mg

114. The specific heat (cal gm^{-1} $°C^{-1}$) of soil organic fraction is

a. 0.25
b. 0.18
c. 0.20
d. 0.46

115. If refers to the difference between moisture contents of a soil at its upper plastic limit and lower plastic limit. Different soils are characterized by a number or index of plasticity.

a. Plastic index
b. Non-plastic
c. Specific plastic
d. liquid limit

116. Atterberg, suggested three values namely upper plastic limit, lower plastic limit and plasticity number or plasticity index and so they are called

a. Site index
b. Non-plastic
c. Power limit
d. Atterberg limits

117. The tensions of the water at the lower plastic limit and upper plastic limit are equivalent to pF values as

a. 2.0 and 1.0
b. 3.0 and 2.0
c. 3.0 and 0.5
d. 1.5 and 0.5

118. The heat flow in a soil through conduction is described by

a. Fourier's low
b. Fick's low
c. Poiseuille low
d. Darcy's law

119. Capillary porosity is determined by finding the volume of water retained in soil at

a. 10 cm tension
b. 20 cm tension
c. 30 cm tension
d. 50 cm tension

120. The thickness of the soil crust increase with the increase in

a. Clay
b. Organic matter
c. Silt
d. Soil aggregate

121. Higher unsaturated hydraulic conductivity of puddled soils in comparison to unpuddled soils is due to

a. Large macro porosity
b. Large total porosity
c. Large micro porosity
d. Higher soil water retention

122. Et (crop) =Et_0 x Kc, where Kc is

a. Climatic factor
b. Day length ratio
c. Crop factor
d. Evaporation factor

123. High zeta potential of a clay colloidal system represents the state of

a. Low swelling
b. High swelling
c. Low viscosity
d. Low plasticity

124. Which soil moisture potential is negative in nature

a. Pressure potential
b. Matric potential
c. Pressure and matric potential
d. Matric and osmotic potential

125. Choose the incorrect statement in given option

a. Bulk density and porosity are inversely related.
b. Compacted soil has lower bulk density of soil.
c. Tillage layer density is lower in ploughed than unploughed.
d. Application of organic matter decreased the bulk density of soil.

126. Which of the following role, the organic matter dose not have

a. Improves soil aggregation
b. Supplies some essential nutrients to soil
c. Favours soil erosion
d. Increases water holding capacity in soil

127. Montmorillonite minerals have external surface area (m^2/g) of

a. 80-150
b. 150-180
c. 180-250
d. 250-300

128. Capillary porosity is given by the formula

a. 1-Field capacity
b. Field capacity x Bulk density
c. Field capacity x Particle density
d. (1- Bulk density/Field capacity)x100

129. The volumetric heat capacity of a wet soil (C_v) is expressed by

a. $C_v = P_b (1+\theta_m) C_p$
b. $C_v = P_b (1-\theta_m) C_p$
c. $C_v = P_b (\theta_m-1) C_p$
d. $C_v = [(1-\theta_m)/P_b] C_p$

where, C_p=heat capacity of soil on mass basis, θ_m=soil water content on mass basis, P_b=bulk density

130. There are number of fundamental assumptions upon which the validity of Stokes' law is based. Which one of the assumptions given below is not correct

a. The particles must be large in comparison to liquid molecules
b. Particles must not be rigid and smooth
c. There must be no slipping between the particle and the liquid
d. The velocity of the fall must not exceed a certain critical value so that the viscosity of the liquid remains the only resistance to the fall of the particles

131. In neutron moisture meter, source of fast neutrons is/are

a. Radium
b. Americium - beryllium
c. Both (a) and (b)
d. Boron

132. Increased bulk density of a soil affects its degree of saturation at a given moisture content by

a. Decreasing it
b. Increasing it
c. Both A and B
d. Does not affect it

133. Moisture retention by soil at low tension/suction is primarily governed by

a. Pore space
b. Pore geometry
c. Void ratio
d. None of these

134. The tension of the water at the upper plastic limit is equivalent to pF value

a. 0.5
b. 1.0
c. 2.0
d. 2.8 - 3.3

135. Water retention in soil is highest when the dominating soil particle is

a. Clay
b. Silt
c. Humus
d. Sand

136. Soil water diffusivity has the units of

a. cm/s
b. cm^2
c. s/cm
d. cm^2/s

137. pF at wilting point is equal to

a. 2.5
b. 4.8
c. 4.2
d. 1.5

138. Percent pore space of soil having bulk density (Pb) and particle density (Ps) is given by

a. (1- Ps/Pb) x 100
b. (1- Pb/Ps) x 100
c. (Pb-Ps) x 100
d. (Ps-Pb) x 100

139. Soil crusting could be reduced by application of

a. Fertilizer
b. Amendment
c. Farm yard manure
d. Surface mulch of plant residues

140. A change in pF value from 3 to 4 results

a. 1
b. 10
c. 100
d. 1000

141. In soil colour determination by Munsell soil colour chart hue is related to

a. Wavelength of light
b. Relative purity or strength of spectral colour
c. Degree of lightness or darkness of a colour
d. Intensity of light.

142. Fractionation of silt and clay fractions of soil particles is done by using

a. Darcy's law
b. Stoke's law
c. Poiseuille's law
d. Fick's law

143. In mechanical analysis, dispersion of soil particles is achieved by shaking the soil with

a. Na_2CO_3 (1N)
b. Na_2SO_4 (1N)
c. NaOH (1N)
d. NaCl (1N)

144. The bulk density of soil affects the thermal conductivity of soils as it

a. Decreases it
b. Does not affect
c. Increases it
d. May increase or decrease it

145. The energy released when soil solids are hydrated is known as

a. Heat of wetting
b. Heat of solution
c. Heat of hydration
d. Heat of sublimation

146. Soil particles which feel sticky and form ribbon without any side crack are known as

a. coarse sand
b. fine sand
c. silt
d. clay

147. The method of soil structure characterization and evaluation that was described by Yodder is:

a. Stability index
b. Wet sieving
c. Dry sieving
d. Mean weight diameter

148. The TDR system of measuring soil moisture at different depths is based on the concept of

a. Neutron scattering
b. Hydraulic resistance
c. Dielectric constant
d. Electrical resistance

149. Crops conducive for the formation of well structured soils :

a. Grasses
b. Cereals
c. Millets
d. Oilseeds

150. Which soil would exhibit a greater hydraulic conductivity?

a. Sand
b. Loam
c. Clay loam
d. Sandy loam

151. Increase in soil temperature

a. Increases hydraulic conductivity
b. Decreases hydraulic conductivity
c. Has no effect on hydraulic conductivity
d. None of these

152. The soil crusts with multiple layers are referred to as

a. Sedimentary crusts
b. Structural crusts
c. Biological crusts
d. Physical crusts

153. If the same amount of water is applied over the surface to wet different dry soils, the depth of wetting that results will be the deepest if the texture of the soil is

a. Silt loam
b. Loam.
c. Sandy loam
d. Clay loam

154. Soils containing almost equal proportion of sand, silt and clay particles are said to be

a. Sandy soils
b. Loamy soils
c. Silty soils
d. Clayey soils

155. The major gas responsible for greenhouse effect is

a. Water vapor
b. Carbon dioxide
c. Methane
d. Ozone

156. Water stable aggregates in soil increase by addition of

a. N- fertilizers
b. P- fertilizers
c. K- fertilizers
d. Micronutrients

157. The long wave radiation emitted by the earth fall in the wavelength range of

a. 4-100 μm
b. 5-100 μm
c. 100-400 μm
d. 200-400 μm

158. Water movement by capillary rise from a water table will rise the highest for which texture

a. Sand
b. Silt loam
c. Clay loam
d. Loam

159. Soil particles which feel buttery when squeezed between thumb and forefinger are known as

a. Coarse sand
b. Fine sand
c. Silt
d. Clay

160. Saturated hydraulic conductivity generally increases, with

a. Increase in upper plastic limit
b. Decrease in upper plastic limit
c. Decrease in lower plastic limit
d. Increase in lower plastic limit

161. Soils containing at least 70% sand and 15% or less clay are said to be

a. Sandy soils
b. Loamy soils
c. Silty soils
d. Clayey soils

162. Bench terraces are constructed for the areas having land slope above :

a. 10%
b. 15%
c. 20 %
d. 25 %

163. The most important cementing agent in laterite soil is

a. Organic matter
b. Calcium carbonate
c. Hydrous oxides of iron and aluminium
d. Clay particles

164. The highest specific surface is noticed in

a. Kaolinite
b. Montmorillonite
c. Illite
d. Chlorite

165. The size of colloidal clay particles are

a. <0.002 mm
b. <0.001mm
c. < 0.0002 mm
d. < 0.0001 mm

166. Munsell colour chart contains about ..... different coloured papers or chips

a. 100
b. 125
c. 150
d. 175

167. How much quantity of total incoming solar radiation is absorbed by the soil and can be used to warm it

a. 10%
b. 20%
c. 42 %
d. 50 %

168. The most important cementing and stabilizing agent in the formation of aggregates among the following is

a. Clay particle
b. Calcium carbonate
c. Microbial gum
d. Humus

169. When a wetting front moving through a clay soil reaches a layer of sand, the rate of movement of the wetting front will

a. Increase
b. Decrease
c. First increase then decrease
d. Remains the same

170. Permafrost is a layer where soil temperature is always

a. 0 °C
b. ~0°C
c. 0 to 100°C
d. 5-10 °C

171. The pipette method for mechanical analysis was first developed by

a. A.H.M. Anderson
b. G.J. Bouyoucos
c. L.D. Bayer
d. G.W. Rabinson

172. The structure of a normal intensive cultivated soil is degraded due to loss of

a. Plant nutrients
b. Soluble salts
c. Organic matter
d. Organic matter plus exchangeable bases

173. The specific surface area of Kaolinite clay is in the range of

a. 7-30 m^2/g
b. 20-50 m^2/g
c. 50-80 m^2/g
d. 1-5 m^2/g

174. Mechanical analysis by hydrometer method was first developed by

a. G.W. Rabinson
b. G.J. Bouyoucos
c. J.R. Philips
d. D. Hillel

175. Relative proportions of sand, silt and clay in soil is called

a. Pedon
b. Soil texture
c. Soil structure
d. Soil complex

176. The natural aggregates in soil are

a. Clod
b. Fragment
c. Ped
d. Mottle

177.The number of classification of soil separates by USDA & ISSS system is

a. Four
b. Six
c. Seven
d. Eight

178. In unsaturated soil conditions, hydraulic conductivity is a function

a. Pressure Potential
b. Matric Potential
c. Osmatic Potential
d. Gravitational Potential

179. The number of textural classes as per the textural triangle is

a. 10
b. 8
c. 12
d. 16

180. The strength of soil crust is measured with the help of

a. Volt meter
b. Radiometer
c. Permeameter
d. Penetrometer

181. Which law is used for studies of water flow

a. Ohm's Law
b. Fick's Law
c. Fourier's Law
d. Darcy's Law

182. The most desirable type of soil structure for field crops is

a. Plate like
b. Block like
c. Prism like
d. Spheroidal

183. Soil aeration limits plants growth when per cent pores filled with water are more than

a. 50 - 60%
b. 60 - 70%
c. 70 - 80%
d. 80 - 90%

184. Which is the upper limit of soil water availability to plants

a. Field capacity
b. Maximum water holding capacity
c. Permanent wilting point
d. Saturation point

185. The external surface area of 1 g of colloidal clay is at least times that of 1 g of coarse sand

a. 10
b. 100
c. 1000
d. 10000

186. The latent heat of vaporization (Cal/g) of water is

a. 80
b. 110
c. 340
d. 540

187. Soil water potential is measured by

a. Tensiometer
b. Penetrometer
c. Pycnometer
d. Hydrometer

188. Which one of the following sub type structure is highly porous?

a. Platy
b. Blocky
c. Granular
d. Crumby

189. Which soil has lowest quantity of available water

a. Clay
b. Sand
c. Clay loam
d. Silt loam

190. Aggregation of a soil can be expressed by

a. Mean weight diameter (MWD)
b. Coefficient of linear extensibility (COLE)
c. Uniformity coefficient (UC)
d. Relative compaction (RC)

191. Water has maximum density at

a. 0°C
b. 100°C
c. 4°C
d. 25°C

192. In Munsell colour chart, the dominant spectral colour is represented by

a. Hue
b. Value
c. Chroma
d. Grade

193. The ratio of volume of water to volume of solid is stated as

a. Volume wetness
b. Degree of saturation
c. Mass wetness
d. None of these

194. Wet sieving technique for soil structure evaluation was first developed by

a. A.F. Tiulin b. G.G. Stokes

c. R.E. Yoder d. W.D. Kemper

195. Bulk density of a soil increases due to

a. Compaction b. Plowing

c. Organic matter addition d. Tank silt addition

196. Particle density of soils can be determined by

a. Hydrometer b. Pycnometer

c. Penetrometer d. Infiltrometer

197. Very poor soil structure is found in

a. Saline soils b. Acid soils

c. Sodic soils d. Saline sodic soils

198. The main process of concentration of free silica in soil is

a. Laterization b. Podzolization

c. Calcification d. Gleization

199. Mull soils develop where the climate is

a. Cool b. Hot

c. Hot and humid d. Moderate

200. Particle density of a soil remains unchanged by

a. Change in organic matter content of the soil

b. Repeated tillage operations

c. Change in mineral particles of soils

d. Increase in soil depth

201. Which is phenomenon is not nuclear phenomenon

a. X-ray emission b. α-particle emission

c. β- particle emission d. Gamma particle emission

202. Dielectric constant of water is

a. 40 b. 80

c. 120 d. 160

203. The process of conversion of solid to vapour without becoming liquid is known as

a. Sublimation
b. Solubilization
c. Vaporization
d. Ionization

204. The diameter of micro pores is

a. < 0.06 mm
b. < 0.8 mm
c. < 0.08 mm
d. < 0.6mm

205. Term fabric was introduced in to soil micro morphology in the year 1938 by

a. G Stoops
b. P Buflock
c. Kubing
d. U Basel

206. The major cementing agent in soil aggregate formation is

a. Lipids
b. Proteins
c. Polysaccharides
d. Mineral nutrients

207. In international pipette method, H_2O_2 is used to remove which of the following

a. $CaCO_3$
b. Iron oxide
c. Organic matter
d. Aluminium oxide

208. The ratio of total volume of pore space to the total volume of soil is called

a. Void ratio
b. Porosity
c. Non capillary porosity
d. Capillary porosity

209. The ratio of total volume of pore space to the total volume of soil solid is called

a. Porosity
b. Capillary porosity
c. Non capillary porosity
d. Void ratio

210. 'Aeolian' deposits are formed by the transportation of materials through

a. Wind
b. Ice
c. Sea
d. Running water

211. The ratio of volume of air to volume of soil is termed as

a. Porosity
b. Air filled porosity
c. Aeration porosity
d. None of-these

212. Broken peds are called as

a. Clods
b. Peds
c. Fragments
d. Aggregates

213. According to Fick's law, diffusive flux of a nutrient is proportional to its

a. Thickness of the plane
b. Concentration gradient
c. Molecular weight of the gas
d. Atomic size of the gas

214. Average annual loss of nutrients due to erosion is

a. 2.4 million tonnes
b. 4.4 million tonnes
c. 8.4 million tonnes
d. 6.4 million tonnes

215. Which law is used in studies of heat flow

a. Darcy's Law
b. Fick's Law
c. Fourier's Law
d. Ohm's Law

216. Maximum swelling occurs in soil containing

a. Vermiculite
b. Montmorillonite
c. Beidellite
d. Illite

217. The saturation extract of a soil has EC 2.0 dS/m. What would be its osmotic potential (in bars)

a. -0.18
b. -0.36
c. -0.72
d. -1.08

218. Total porosity is more in

a. Sandy soil
b. Clayey soil
c. Loamy soil
d. Silty soil

219. Which of the following is known as physically dry soil?

a. Saline soil
b. Clay soil
c. Sandy soil
d. Calcareous soil

220. Chemically most active soil separate is

a. Sand
b. Silt
c. Clay
d. Gravel

221. Capillary fringe in soil refers to height of water rise in a capillary

a. Saturated zone above the water table
b. Water in capillary pores
c. Depth of water table
d. Below to saturated zone

222. The best agricultural soil structure is

a. Crumb
b. Platy
c. Columnar
d. Blocky

223. The optimum condition for tillage is

a. Harsh
b. Friable
c. Plastic
d. Consistency

224. One acre of soil to a depth of 6 inch weight about

a. 2.25×10^6 kg
b. 2×10^6 kg
c. 2×10^6 lb
d. 2.25×10^6 lb

225. Soils are easily puddled at

a. Friable
b. Plastic
c. Sticky
d. Consistency

226. The ground water table depth is measured by

a. Gypsum block
b. Piezometer
c. Neutron probe
d. Odometer

227. Which types soil structure found common in the B horizon, particularly in humid regions

a. Platelike
b. Prismlike
c. Blocklike
d. Crumby

228. One bar equal to

a. 100 kPa
b. 0.99 atm
c. 14.5 psi
d. All of these

229. The volumetric moisture content of soil can be computed by multiplying the gravimetric moisture content with

a. Particle density
b. Bulk density
c. Porosity
d. Void ratio

230. The ratio of volume of soil air to volume of soil is called as

a. Porosity
b. Air filled capacity
c. Aeration porosity
d. Bulk density

231. The minimum water content at which soils are puddled is

a. Flocculation limit
b. Upper plastic limit
c. Lower plastic limit
d. Shrinkage limit

232. In remote sensing, which type of radiation is utilized for photography

a. Violet
b. Far red light
c. Infra red
d. Green light

233. Soils are friable at

a. Flocculation limit
b. Upper plastic limit
c. Lower plastic limit
d. Shrinkage limit

234. Specific heat of organic matter is

a. 0.11 Cal/g
b. 0.2 Cal/g
c. 1.0 Cal/g
d. 0.44 Cal/g

235. Choose the correct statement in given option

a. The thermal conductivity appears as a proportionality coefficient in fick's equation.
b. The thermal conductivity decreases with soil compaction
c. The thermal conductivity increases with reduction in particle size
d. The thermal conductivity increases with the increase in soil moisture content of dry soil

236. Choose the correct statement in given option
   a. The potential plasticity and cohesion are always high in sandy soils
   b. A dry pulverized soil cannot be compacted to high density
   c. Stress can be defined as the change in size or shape of a body caused by forces acting on it
   d. Strain may be defined as the force inside a body restricting change in size or shape
237. Choose the correct statement in given option
   a. In soil weight determination particle density is helpful
   b. To understand physical behaviour of soil, particle density is more important than bulk density
   c. Specific gravity and particle density of soil are mostly same.
   d. Soil property does not change easily
238. Choose the correct statement in given option
   a. The clay of saline soils is often deflocculated.
   b. The reclamation of sodic soils is easier than saline soils.
   c. Coarse textured soils are generally more affected by exchangeable sodium than fine textured soils
   d. The degree of swelling is related to the specific surface of soils
239. Thermal conductivity of which type of soil is higher
   a. Clayey soils
   b. Silty loam soils
   c. Sandy soils
   d. Clay loam soils
240. Pressure membrane apparatus is used to measure matric potential moisture content relations at potential values as low as
   a. 1 bar
   b. 10 bar
   c. 100 bar
   d. 1000 bar
241. Sticking of two different nature of particles is called
   a. Flocculation
   b. Cohesion
   c. Deflocculation
   d. Adhesion
242. The size of silt particles in ISSS system is:
   a. 2 mm to 0.2 mm
   b. 0.2 mm to 0.02 mm
   c. 0.02 mm to 0.002 mm
   d. <0.002 mm

243. The major cation causing deflocculating of soil is:

a. Na^+
b. K^+
c. Ca^+
d. Mg^+

244. Weight of one hectare furrow slice of root zone depth (15 cm) of soil is about:

a. 2×10^6 kg
b. 1.5×10^6 kg
c. 20×10^6 kg
d. 15×10^6 kg

245. Neutron probe is used for determination of

a. Soil moisture
b. Leaf moisture
c. Soil texture
d. Soil temperature

246. Stokes' law suggests the rate of fall of particle is directly proportional to the square of :

a. Density of the particle
b. Density of the fluid
c. Viscosity of the medium
d. Radius of the particle

247. Rocks or soil debris accumulate at the foot of slope due to gravity is known as

a. Lacustrine
b. Alluvial
c. Fluvial
d. Colluvial

248. The working range of soil tensiometer is :

a. 0-1.0 bar
b. 0-0.85 bar
c. 0-1.85 bar
d. 0-1.5 bar

249. In water erosion if rill is too big to be crossed by farm equipment, it is called

a. Sheet erosion
b. Plate erosion
c. Gully erosion
d. Big rill erosion

250. Major plant usable water in soils is :

a. Capillary
b. Hygroscopic water
c. Gravitational water
d. Lattice water

251. Atterberg constant is related to the soil property.

a. Structure
b. Elasticity
c. Plasticity
d. None of these

252. Process of podzolization takes place in

a. Warm humid climate
b. Arid climate
c. Cold humid climate
d. Semi-arid climate

253. The total pore space is more in

a. Sandy soil
b. Silty loam soil
c. Loam soil
d. Clay loam soil

254. The property that is not affected by the amount of organic matter in the soil is

a. Cation exchange capacity
b. Soil texture
c. Biological activity
d. Soil structure

255. Measurement of soil-moisture by Time Domain Reflectometer (TDR) is based on

a. Neutron scattering
b. Dielectric constant of water
c. Electrical resistance of soil
d. Viscosity of water

256. Rain drop splash and surface flow cause

a. Splash erosion
b. Gully erosion
c. Rill erosion
d. Sheet erosion

257. Stokes' law applies to the settling of spheres in

a. Stationary liquid and constant temperature
b. Stationary liquid and variable temperature
c. Mobile liquid and constant temperature
d. Mobile liquid and variable temperature

258. Specific heat of quartz among the different soil constituents is

a. Lowest
b. Highest
c. Similar to other constituents
d. Variable

259. Different forms of soil consistency were suggested by

a. E.W. Russel
b. L.D. Bayer
c. Atterberg
d. A. Keen

260. The water holding capacity (WHC) of humus on a mass basis is times that of the silicate clays

a. 2- 3
b. 4 - 5
c. 9- 10
d. 100

261. The diameter of water molecules is

a. 2.24 A
b. 2.44 A
c. 2.64 A
d. 2.84A

262. The pF value of saturated soil and oven dry soil is

a. 0 and 14
b. 0 and 7
c. 1 and 14
d. 1 and 7

263. Water holding capacity among the following soil is maximum in

a. Loamy sands
b. Sandy loam
c. Loam
d. Silty loam

264. Dispersing agent used in particle size analysis of soil is :

a. Sodium carbonate
b. Sodium thiosulphate
c. Sodium hexametaphosphate
d. Sodium phosphate

265. Fick's law govern the mechanism of

a. Mass flow
b. Diffusion
c. Capillary movement
d. Laminar flow

266. Under normal field condition, the soil water potential is

a. Positive
b. Negative
c. Zero
d. Unity

267. Pressure plate apparatus is used for measurement of soil moisture tension up to :

a. 0.33 bars
b. 31 bars
c. 4.2 bars
d. 15 bars

268. Water tightly held in thin films around soil particles is

a. Hygroscopic water
b. Gravitational water
c. Capillary water
d. Heavy water

269. Percent solid space in soil is expressed by :

a. Bulk density/Parade density x 100
b. Particle density/Bulk density x 100
c. (Particle density-Bulk density) x 100
d. (Bulk density-Particle density) x 100

270. Tensiometer is used to measure

a. Gravitational potential
b. Matric potential
c. Osmotic potential
d. Pressure potential

271. Total pore space of a soil gives an idea about:

a. Maximum water holding capacity
b. Field capacity
c. Capillary water
d. Gravitational water

272. Scattering of detached soil particles by the impact of rain drops is called

a. Ravines
b. Gully erosion
c. Sheet erosion
d. Splash erosion

273. In Universal soil loss equation, A= RKLSCP, K denotes

a. Soil conservation practices
b. Soil roughness
c. Soil erodability
d. Rainfall factor

274. Of the following, which one is important with regards to soil erosion by water ?

a. Intensity of rainfall
b. Duration of rainfall
c. Amount of rainfall
d. Frequency of rainfall

275. Which type of erosion is the most highly visible ?

a. Gully erosion
b. Wind erosion
c. Sheet erosion
d. Splash erosion

276. Driving force for unsaturated flow in soil is:

a. Gravitational force
b. Matric suction
c. Osmotic suction
d. Solute potential

277. The plant available water retained in soil is between

a. 0 to 3 bar
b. -1 to-1/3 bar
c. 15 to 30 bar
d. -1/3 to-15 bar

278. The difference in the water contents between upper plastic limit and lower plastic limit is called

a. Plasticity index
b. Viscosity index
c. Swelling index
d. Shrinkage index

279. Swelling and shrinkage properties of soil can be expressed by

a. Coefficient of viscosity
b. Coefficient of expansion (COEP)
c. Coefficient of linear extensibility (COLE)
d. Coefficient of vertical expansion (COVE)

280. At field capacity the moisture is held with soil water potential

a. < 1 bar
b. < 1/3 bar
c. < 0.5 bar
d. > 1/3 bar

281. In gravimetric method of soil moisture measurement, the soil is oven dried at

a. 90°C - 100 °C
b. 105°C - 110 °C
c. 115°C - 120 °C
d. 125°C- 130 °C

282. Which of the following method of measurement of soil moisture is considered as check for all other methods

a. Electrical conductivity method
b. Tensiometer method
c. Gravimetric method
d. Neutron probe method

283. Estimate the percent moisture in a given : Weight of moist soil + box = 200 gm, Weight of oven dry soil + box = 160 gm and Weight of empty box = 50 gm

a. 22.22% b. 40.60%

c. 36.36% d. 36.84%

284. Available water in soil refers to that water which is held between

a. Maximum water holding capacity and field capacity
b. Field capacity and hygroscopic coefficient
c. Wilting point and hygroscopic coefficient
d. Field capacity and wilting coefficient

285. More accurate analysis of soil texture can be done by

a. Hydrometer method b. International pipette method

c. Aggregate method d. Yoder method

286. Neutron moisture probe estimates moisture in soil by detecting molecules of :

a. Water b. Oxygen

c. Hydrogen d. Carbon

287. At field capacity the pF value of heavy soils is

a. 1.0 b. 2.5

c. 4.2 d. 4.5.

288. The soil erodibility factor (K) indicates the inherent erodibility of a soil gives an indication of the soil loss from a unit plot m long with a 9 % slope under continuous fallow culture

a. 22 b. 40

c. 62 d. 72

289. Land capability class restricted to recreation, wildlife, water supply, or aesthetic purposes is

a. I b. VI

c. VII d. VIII

290. Moisture content of soil at a point where the soil water mass flows under an applied force and fails to retain its shape is called

a. Plastic index
b. Non-plastic
c. Lower plastic limit
d. Upper plastic limit (liquid limit)

291. Presence of silica, lime and other salts in soil, the colour of soil appears like

a. Dark coloured
b. Brown coloured
c. Red coloured
d. Light coloured

292. Presence of organic matter and iron oxides in the soil, the colour of soil appears like

a. Red colour
b. Brown colour
c. Dark colour
d. Yellow, red colour

293. The relation between osmatic pressure (bar or atmosphere) and electrical conductivity for salt mixture is OP = 0.36 X EC, the EC is expressed as in MKS system

a. dS/m
b. mhos/cm
c. μ mhos/cm
d. d mhos/cm

294. With the increase in organic matter of the given soil the Particle Density (P.D.)

a. Does not change (No change)
b. Increases
c. First decreases and then increases
d. Decreases

295. Specific heat may be defined as the amount of heat required to raise the temperature of one gram substance by

a. 1°F
b. 1°K
c. 1°C
d. 0.1°F

296. If bulk density of soil is 1.5 Mg m^3, find the weight of 0-10 cm depth of one hector soil in kg.

a. 2.2×10^6 kg
b. 1.5×10^6 kg
c. 2.25×10^6 kg
d. 3.0×10^6 kg

297. Particle density (g cm^{-3})=

a. Weight of moist soil/volume of soil
b. Weight of dry soil/volume soil
c. Weight of dry soil/volume of soil particles
d. Volume of soil/weight of moist soil

298. Soil available water content (%) at filed capacity is maximum/highest in

a. Loam soil
b. Silt loam soil
c. Clay loam soil
d. Clayey soil

299. The relationship among the Field Capacity (FC), Permanent Wilting Point (PWP) and Available Water Capacity (AWC) is

a. FC = AWC - PWP
b. AWC = FC - PWP
c. PWP = FC -AWC
d. AWC = PWP- FC

300. The Nomenclature of soil is typic hyper thermic ustipssment is then the ustic stand for

a. Temperature regime
b. Textural class
c. Moisture regime
d. Mineral class

301. Anion responsible for soil aggregation is:

a. Nitrate
b. Phosphate
c. Sulphate
d. Molybdate

302. Where the MAST is 22°C or higher and the difference between MSST and MWST is 6°C or more, the soil temperature regime is termed as

a. Hyperthermic
b. Isohyperthermic
c. Mesic
d. Thermic

303. According to Darcy's law, flux is

a. Proportional to hydraulic head
b. Proportional to hydraulic gradient
c. Equal to hydraulic head
d. Equal to hydraulic gradient

304. In water the two hydrogen atoms are attached with the oxygen atom at angle

a. 90°C
b. 100°C
c. 105°C
d. 110°C

305.The density of water is maximum at

a. 0°C
b. 4°C
c. 20°C
d. 25°C

306.The latent heat of vaporization of water at 20°C is

a. 537 cal/gm
b. 540 cal/gin
c. 590 cal/gm
d. 100 cal/gm

307. The dielectric constant of pure water at 25°C is

a. 72.7
b. 78.5
c. 80.2
d. 90.5

308. The process of water entry into soil through the surface may be either downward or lateral or both is called as :

a. Percolation
b. Infiltration
c. Seepage
d. Leaching

309. What is the unit of hydraulic conductivity

a. cm sec^{-2}
b. m sec^{-1}
c. cm sec^{-1}
d. m sec^{-2}

310. The potential results from the presence of solutes in soil water is called as:

a. Pressure Potential
b. Osmotic Potential
c. Metric Potential
d. Gravitational Potential

311. What is the soil water potential at plant available water ?

a. 10-30 kPa
b. 33 kPa
c. 1500 kPa
d. 33-1500 kPa

312. Arrange the following soil texture in the order of decreasing water retention

a. Loam>Clay>Sandy
b. Sandy>Loam>Clay
c. Sandy>Clay>Loam
d. Clay>Loam>Sandy

313. Identify particle sizes by least to greatest or fine to coarse

a. Sand,-Silt, Clay
b. Clay, Sand, Silt
c. Silt, Sand, Clay
d. Clay, Silt, Sand

314. Soil water held at 1/3 bar is called

a. Field capacity
b. Available water .
c. Wilting point
d. Permanent wilting point

315. The concept of pF was first introduced by

a. E. Buckingham
b. C.E. Marshall
c. R.K. Sehofield
d. L. A. Richards

316. In neutron moisture meter the thermalized neurons are

a. Slow neutrons
b. Fast neutrons
c. Heated neutrons
d. Heat resistant neutrons

317. Which soil among the following has greatest maximum water holding capacity?

a. Sandy loam
b. Silt loam
c. Clay loam
d. Clay

318. Which soil among the following has greatest field capacity?

a. Clay loam
b. Silt loam
c. Loam
d. Sandy loam

319. The matric potential is also called as

a. Submergence potential
b. Suction potential
c. Gravitational potential
d. Osmotic potential

320. .... is considered as first step of soil erosion caused due to water

a. Gully erosion
b. Rill erosion
c. Slpash erosion
d. Sheet erosion

321. Capillary rise of water is due to the forces of

a. Adhesion
b. Cohesion
c. Adhesion and cohesion
d. Atmospheric pressure

322. If 5 cm rainfall occurs on a hectare land during one hour and if 50 percent of rain water is lost as run-off from the field, then value of water collected in one hectare of field will be:

a. 150 m^3
b. 250 m^3
c. 350 m^3
d. 500 m^3

323. Maximum wind erosion is recorded in the season of

a. Winter
b. Rainy
c. Summer
d. Spring

324. The main reason for sheet erosion is

a. Wind
b. Rainfall
c. Temperature
d. Organic matter

325. Wind erosion is the highest in the State of India

a. Kerala
b. Jharkhand
c. Madhya Pradesh
d. Rajasthan

326. Hydrometer is used to determine

a. Soil moisture content
b. Soil aggregate size
c. Soil hydraulic conductivity
d. Soil texture

327. Strip cropping involves growing of

a. Row crops with high density
b. Strips of high density row crops
c. Alternate strips of row cover crops
d. Alternate strips of row and cover crops along the contour

328. Soil pF is a negative log of

a. H^+ ion concentration
b. Flouride ion concentration
c. cm of the mercury column
d. Flouride ion activity

329. What is bulk density of 100 cm^3 soil, if its oven dry weight is 135 grams

a. 13,500 g/cm^3 b. 0.75 g/cm^3

c. 1.35 g/cm^3 d. 1.53 g/cm^3

330. Which law follows the heat flow through soil

a. Darcy's law b. Fourier's law

c. Ohm's law d. Fick's law

331. Which law follows water flow through soil?

a. Darcy's law b. Fourier's law

c. Ohm's law d. Fick's law

332. Silt content of a soil is a good indicator of

a. Siltation of reservoirs b. Structure of soil

c. Erodibility of soil d. Fertility of soil

333.Which is not associated with wind erosion?

a. Siltation b. Overland flow

c. Surface creeping d. Suspension

334. How many numbers of classes consist in land capability classification?

a. Eight b. Four

c. Six d. Ten

335. In land capability classification, the lands falling under Class V to Class VIII are

a. Suitable for Cultivation

b. Unsuitable for cultivation

c. Suitable for only wildlife conservation

d. Only for grassland farming

336. Movement of soil particles in series of bounces or jumps along surface of ground is termed as

a. Surface creep b. Suspension

c. Saltation d. Threshold velocity

337. For normal growth of soil organisms the ODR values should be at least

a. 10×10^{-8} gm/cm^2-min
b. 40×10^{-8} gm/cm^2-min
c. 50×10^{-8} gm/cm^2-min
d. 100×10^{-8} gm/cm^2-min

338. In Darc's law, the hydraulic conductivity (K) has the physical dimensions of

a. $ML^{-1}T^2$
b. $M^{-1}L^3T$
c. L^3T^{-1}
d. LT^{-1}

339. Mechanical methods of erosion control can be viewed as an attempt to

a. Increase in soil resistance to erosion
b. Control the energy for erosion
c. Increase soil erosivity
d. Decrease rainfall intensity

340. Which of the following properties are described by Phillip'e equation (I=St1/2+Kt)

a. Gas diffusion
b. Soil permeability
c. Thermal diffusion
d. Soil infiltration

341. Which is not matched with conservation tillage

a. Increases soil organic matter content
b. Reduces cost of cultivation
c. Destructs soil aggregation
d. Increase the incidence of plant diseases from crop residues

342. Dry land farming is the practice of crap production receiving the annual rainfall of

a. <800mm
b. <1000mm
c. <900mm
d. <1100mm

343. Plasticity of soil can be determined by using

a. Pycnometer
b. Buoyoucos hydrometer
c. Atterberg's apparatus
d. Penetrometer

344. In universal soil loss equation, Xa=RKLSCP, R represents....... factor

a. Rainfall erosivity
b. Soil erodibility
c. Slope length
d. Conservation practices

345. According to Stock's law, the terminal velocity of a spherical particle settling under the influence of gravity in a fluid of given density and viscosity is

a. Proportional to the radius
b. Proportional to the square of the radius
c. Equal to the radius
d. Equal to the square of the radius

346. Soil aggregation can be estimated by using

a. Yoder's apparatus
b. MIR
c. Cone penetrometer
d. Atterberg's apparatus

347. In........ structure, peds are arranged in relatively thin horizontal plates

a. Plate like
b. Blocky
c. Prism like
d. Spheroidal

348. Application of organic matter to a fine textured soil results in an increase of its

a. Aeration porosity
b. Bulk density
c. Particle density
d. Capillary porosity

349. The specific heat (cal gm^{-1} $°C^{-1}$) of soil mineral fraction is

a. 0.25
b. 0.18
c. 0.20
d. 0.46

350. The specific heat (cal $gm^{-1}°C^{-1}$) of soil solid is

a. 0.25
b. 0.18
c. 0.20
d. 0.46

351. The colour notation of Class II land is

a. Yellow
b. Red
c. Orange
d. Brown

352. Hue variable in Munsell Colour Chart describes which of the followings

a. Lightness/darkness of the dominant spectral colour

b. Dominant spectral colour

c. Purity of dominant colour

d. Intensity of dominant spectral region

353. In the universal soil loss equation, A=RKLSCP, 'S' stands for

a. Sodium

b. Slope

c. Salts

d. Sand

354. The water balance in soil profile is a net result of processes that do not include

a. Infiltration

b. Percolation

c. Drainage

d. Surface evaporation

355. Routine soil particle size distribution analysis can be easily and quickly done by one of the following methods

a. Day's hydrometer method

b. Bouyoucos hydrometer method

c. International pipette method

d. Sur and Kukal's hydrometer method

356. Flow velocity increases with

a. Increase in hydraulic radius

b. Increase in rain drop size

c. Decrease in hydraulic radius

d. Increase in depth of surface runoff

357. In Munsell soil colour chart hue and value ranges from

a. 0 -8

b. 1- 8

c. 0-10

d. 1-10

358. In Munsell soil colour chart the chroma number ranges from

a. 0-8

b. 1-8

c. 0-10

d. 1-10

359. The basic form of present day USLE initially involved the use of factors such as

a. R, K and S
b. R, K,S and L
c. R and K
d. R, K, S, L, C and P

360. The thermal conductivity of soil varies with soil texture in the order of

a. Sand>loam>clay>peat
b. Loam>clay>sand>peat
c. Peat>clay>loam>sand.
d. Clay>loam>clay>peat

361. The Soil Physicists in our country recently recommended soil matric potential as criterion for scheduling intermittent irrigation to rice crop. They recommended the value of soil matric potential for this purpose as

a. -22kPa
b. 25kPa
c. -16kP
d. 16kPa

362. The moisture equivalent was introduced by

a. Briggs and McLane (1907)
b. Richards and Weaver (1944)
c. Shaw and Bayer (1939)
d. Whitney

363. The ratio of maximum rainfall at a point and to the mean total rainfall in a catchment is called

a. Rainfall aggressiveness
b. Rainfall distribution
c. Distribution coefficient
d. Rainfall Intensity

364. The volume wetness of soil can be derived from which one of the following

a. Mass wetness and bulk density
b. Bulk density and particle density
c. Hydraulic conductivity maximum WHC
d. Particle density and intrinsic permeability

365. The change in apparent specific volume of soil per unit of work expanded in causing such change is called

a. Erodibility
b. Bulk density
c. Puddlability
d. Specific density

366. The Bouyoucos hydrometer method recommends to take hydrometer reading at what respective times for determining silt+clay and clay contents at 68°F

a. 40 seconds and 2 hours
b. 2 minutes and 4 hours
c. 40 secdonds and 3 hours
d. 3 minutes and 2 hours

367. A sample of soil weighs 110 g and contains 75% moisture. If the volume of the soil is 75 cm^3, what is the bulk density of the soil ?

a. 1.80 g/cm^3
b. 1.52 g/cm^3
c. 1.47 g/cm^3
d. 1.28 g/cm^3

368. Given two points (A and B) in soil, one located at a distance (vertical) of 15 cm above and the other 10 cm below a reference level elevation. Find the difference in gravitational potential between the two points

a. 25 cm
b. 5 cm
c. -25 cm
d. -15 cm

369. Among the following, the sequence of textural classes of soil with increasing order of fineness of texture is

a. CL, L, LS, SL, SCL
b. L, CL, SCL, SL, LS
c. SL, SCL, LS, L, CL
d. LS, SL, L, SCL, CL

370. Soil moisture retention, with all other factors remaining the same, is usually in the order of

a. Montmorillonite>Kaolinite>Illite
b. Illite>Kaolinite>Montmorillonite
c. Montmorillonite>Illite>Kaolinite
d. Halyosite>Kaolinite>Illite

371. Withan increase in soil pHfrom 2 to 8, the Eh of the soil

a. Decreases
b. Increases
c. Is not affected
d. May increase or decrease

372. A soil having porosity of 50% contains volumetric water content of 25% its degree of saturation will be

a. 25%
b. 50%
c. 2%
d. 5%

373. The dominant spectral colour in soil indicates

a. Hue
b. Value
c. Chroma
d. Organic matter

374. The soils most prone to structural crusts are those which are higher in

a. Sodium content
b. Calcium content
c. Clay content
d. Silt content

375. The root growth ceases when the ODR drops to

a. 10×10^{-8} g cm^2/min
b. 20×10^{-8} g cm^2/min
c. 30×10^{-8} g cm^2/min
d. 40×10^{-8} g cm^2/min

376. The differential relationship between soil water content and soil matric potential due to wetting and drying cycles is termed as

a. Soil moisture characteristics
b. Soil water desorption
c. Hysteresis
d. Soil water diffusivity

377. Gully texture refers to as

a. Number of total gullies per unit area
b. Length of gullies per unit area
c. Number of first-order gullies per unit area
d. Length of highest order gullies per unit area

378. The physical fertility of the soil depends more on the following physical parameters

a. Infiltration and Drainage
b. Thermal conductivity and diffusivity
c. Soil structure and texture
d. Soil structure and hydraulic conductivity

379. The basic unit of soil structure formed by clay cementation around a microbial colony is termed as

a. Micro aggregate
b. Macro aggregate
c. Domain
d. Aggregate

380. Soil colour determination chroma refers to which of the following ?

a. Dominant spectral colourb.

Lightness or darkness of the colour

c. Purity or strength of colour

d. None of these

381. The upper size limit of clay particle is

a. 0.02 mm b. 0.002 mm

c. 2.0 mm d. 0.2 mm

382. Thermal conductivity of the soil is in which of the following orders

a. Sand > loam > clay > peat b. Sand > peat > loam > clay

c. Loam > sand > peat > clay d. Peat > sand > loam > clay

383. The difference between upper plastic limit and lower plastic limit is called as

a. Plasticity index b. Plasticity number

c. Plasticity coefficient d. COLE

384. Specific heat is higher in

a. Sand b. Silt

c. Clay d. Humus

385. Soil water potential is generally

a. Zero b. Neutral

c. Positive d. Negative

386. When soil is saturated, its moisture content is

a. Equal to field capacity b. Greater than field capacity

c. Less than field capacity d. Equal to PWP

387. The following soils have high osmotic potential

a. Alkaline soils b. Acid soils

c. Calcareous soils d. Saline soils

388. Aggregate stability of soils is determined by

a. Yoder's apparatus
b. Pressure plate apparatus
c. Atterberg's apparatus
d. Penetrometer

389. Most used coagulant is

a. Chlorine
b. Lime
c. Alum
d. Bleaching powder

390. Thermo couple psychrometer is used to measure

a. Soil water potential
b. Hydraulic conductivity
c. Osmotic potential
d. Infiltration

391. High specific heat and high latent heat are the characteristics of

a. Water
b. Soil
c. Mercury
d. Ethanol

392. Available water holding capacity is maximum in

a. Clay loam
b. Silty clay loam
c. Sandy clay loam
d. Loam

393. The ratio of total mass of soil solids to total volume of soil solids is called

a. Particle density
b. Bulk density
c. Porosity
d. Air filled porosity

394. The manifestations of the physical forces of cohesion and adhesion acting within the soil at various moisture contents are designated as

a. Soil consistency
b. Soil plasticity
c. Soil compression
d. Soil compaction

395. When water potential of Cell A is -10 bars arid Cell B is -18 bars, water moves from

a. B to A
b. A to B
c. Does not move
d. Move in both direction

396. Height of rise of water in a capillary tube is directly proportional to

a. Surface tension of water
b. Radius of capillary tube
c. Viscosity of water
d. As in (a) and (c) above

397. Plasticity of soil can be determined by

a. Hydrometer
b. Atterberg's apparatus
c. Pressure Plate Apparatus
d. Tensiometer

398. The relative proportion of sand, silt and clay in a soil is known

a. Soil texture
b. Soil structure
c. Soil clods
d. Soil aggregation

399. Arrangement of soil particles is referred to as

a. Sod texture
b. Soil structure
c. Soil aggregate stability
d. Bulk density

400. The value of Avogardo's number is

a. $6.013x10^{19}$
b. $6.023x10^{21}$
c. $6.023x10^{-23}$
d. $6.023x10^{23}$

401. What is the porosity percentage if the particle density of soil is 2.6 and bulk density is 1.3

a. 40%
b. 50%
c. 55%
d. 65%

402. Thermal diffusivity of a soil having thermal conductivity K and volumetric heat capacity C is given by

a. K/C
b. C/K
c. KC
d. $K/\sqrt{C}$

403. Land area having definite boundary and contributing runoff to a single point is known as

a. Watershed
b. Catchment area
c. Command area
d. A field unit

404. Moisture content at wilting point is highest in

a. Clay loam
b. Silty clay loam
c. Loam
d. Sandy loam

405. The size of micro-watershed is

a. 10,000-20,000 ha
b. 1,000-10,000 ha
c. 100-1,000 ha
d. 10-100 ha

406. Darcy's law is applicable under unsaturated conditions if hydraulic conductivity is expressed as a function of

a. Porosity
b. Hydraulic gradient
c. Hydraulic head
d. Water content

407. In one hector area required water for irrigation of one cm depth to

a. 100 tonnes water
b. 10^6 litres water
c. Both (a) and (b)
d. 10 tonnes water

408. TDR is a method of monitoring

a. Soil wetness
b. Vapour pressure
c. Salt concentration
d. Solar radiation

409. The international pipette method is based on the principle of

a. Decreasing density of soil suspension
b. Sedimentation
c. Silt content of soil
d. Sand content of soil

410. In the universal soil loss equation, factor 's' stands for

a. Slope-length
b. Slope gradient
c. Soil texture
d. Soil erodibility

411.Double rings infiltrometer is used for

a. Maximizing the vertical and horizontal flows
b. Reducing the vertical flow
c. Reducing the upward flow
d. Minimizing the horizontal flow

412. The numerical notation 2.5 YR 5/6 Indicates equivalent soil colour name as

a. Yellow
b. Yellow red
c. Red
d. Red yellow

413. If total soil water potential at a point A in soil is -150 cm and at point B is -200 cm, then

a. Water will flow from point A to B
b. Water will flow from point B to A
c. There will be no flow of water between points A & B
d. More information is needed to draw conclusion on direction of flow of water

414. Match List I with List II

a. Soil bulk density	1. Hydrometer
b. Particle size analysis of soil	2. Picnometer
c. Soil moisture estimation	3. Permeameter
d. Soil moisture retention	4. TDR

415. Match List I (Scientist) with List II (Area of contribution)

a. Mohr	1. Water logging
b. Munsell	2. Soil moisture
c. Atterberg limit	3. Soil colour chart
d. C. F. Clement	4. Relative hardness of minerals

416. Match List I with List II

a. Hydraulic conductivity	1. Stokes' law
b. Thermal conductivity	2. Fourier law
c. Diffusion coefficient	3. Fick's law
d. Terminal velocity	4. Intrinsic permeability
e. Sorptivity	5. Infiltration

417. Match List I with List II

a. Ratio of total volume of soil to the total mass of dry soil
1. Porosity
b. Ratio of total mass of soil solids to the total volume of soil solids
2. Void ratio
c. Ratio of total volume of pore spaces to the total volume of soil
3. Air filled porosity

d. Ratio of total volume of pores to the total volume of soil solids
4. Specific volume
e. Ratio of volume of air space to the total volume of soil
5. Particle density

ANSWERS (Soil Physics)									
1.c	2.c	3.a	4.a	5.d	6.c	7.b	8.a	9.c	10.b
11.a	12.a	13.c	14.c	15.a	16.c	17.a	18.b	19.a	20.d
21.a	22.d	23.a	24.b	25.a	26.b	27.c	28.d	29.d	30.b
31.b	32.b	33.d	34.a	35.a	36.c	37.c	38.d	39.a	40.d
41.b	42.b	43.b	44.a	45.a	46.d	47.b	48.b	49.a	50.a
51.d	52.b	53.b	54.a	55.a	56.c	57.d	58.a	59.c	60.b
61.c	62.c	63.b	64.c	65.d	66.c	67.a	68.d	69.c	70.d
71.b	72.b	73.b	74.b	75.c	76.b	77.a	78.b	79.a	80.b
81.d	82.c	83.b	84.b	85.d	86.a	87.d	88.b	89.c	90.a
91.b	92.c	93.a	94.a	95.a	96.d	97.c	98.a	99.b	100.b
101.a	102.c	103.d	104.c	105.d	106.c	107.b	108.c	109.c	110.c
111.a	112.a	113.a	114.d	115.a	116.d	117.c	118.a	119.d	120.a
121.c	122.c	123.a	124.d	125.b	126.c	127.a	128.b	129.a	130.b
131.c	132.a	133.a	134.a	135.c	136.d	137.c	138.b	139.d	140.b
141.a	142.b	143.c	144.c	145.a	146.d	147.d	148.c	149.a	150.a
151.a	152.b	153.c	154.b	155.a	156.b	157.a	158.c	159.c	160.b
161.a	162.b	163.c	164.b	165.c	166.d	167.a	168.d	169.a	170.a
171.d	172.c	173.a	174.b	175.b	176.c	177.a	178.b	179.c	180.d
181.d	182.d	183.a	184.a	185.c	186.d	187.a	188.d	189.b	190.a
191.c	192.a	193.b	194.a	195.a	196.b	197.c	198.a	199.c	200.b
201.a	202.b	203.a	204.c	205.c	206.c	207.c	208.b	209.d	210.a
211.b	212.c	213.b	214.a	215.c	216.a	217.c	218.b	219.c	220.c
221.a	222.a	223.b	224.c	225.b	226.b	227.c	228.d	229.b	230.b
231.c	232.c	233.d	234.d	235.d	236.b	237.c	238.d	239.c	240.c
241.d	242.c	243.a	244.a	245.a	246.d	247.d	248.b	249.c	250.a
251.c	252.c	253.d	254.b	255.b	256.a	257.a	258.b	259.c	260.b
261.c	262.b	263.d	264.c	265.b	266.b	267.d	268.a	269.a	270.b
271.a	272.d	273.c	274.a	275.a	276.b	277.d	278.a	279.c	280.d
281.b	282.c	283.c	284.d	285.b	286.c	287.b	288.a	289.d	290.d
291.d	292.b	293.a	294.a	295.c	296.c	297.c	298.b	299.b	300.a
301.b	302.a	303.b	304.c	305.b	306.c	307.b	308.b	309.c	310.b
311.d	312.d	313.d	314.a	315.c	316.a	317.d	318.a	319.b	320.c
321.c	322.b	323.d	324.b	325.d	326.d	327.d	328.c	329.c	330.b
331.a	332.d	333.b	334.a	335.b	336.c	337.b	338.d	339.b	340.d
341.c	342.a	343.c	344.a	345.b	346.a	347.a	348.a	349.b	350.c
351.a	352.b	353.b	354.d	355.b	356.c	357.c	358.a	359.d	360.a

ANSWERS (Soil Physics)									
361.d	362.a	363.a	364.a	365.c	366.a	367.d	368.b	369.d	370.c
371.a	372.b	373.a	374.c	375.b	376.c	377.c	378.c	379.a	380.c
381.b	382.a	383.a	384.d	385.d	386.b	387.d	388.a	389.b	390.a
391.a	392.a	393.a	394.a	395.b	396.a	397.b	398.a	399.b	400.d
401.b	402.a	403.a	404.b	405.c	406.d	407.c	408.a	409.b	410.b
411.d	412.b	413.a							
414. a.3 b.1 c.4 d.2					415. a.4 b.3 c.2 d.1				
416. a.1. 2 c. 3 d.4 e.5					417. a.4 b.5 c.1 d.2 e.3				

6

Pedology

1. A mineral that forms an original component of rock is called

   a. Primary b. Secondary

   c. Essential d. Accessory

2. The science of soil where attention is given to the soil-plant relationship would be known as:

   a. Geology b. Petrology

   c. Edaphology d. Pedology

3. The saline and alkali soils fall in which order

   a. Entisols b. Alfisols

   c. Mollisols d. Aridisols

4. Who is called father of Pedology

   a. N. C. Brady b. H. Jenny

   c. V. V. Dokuchaev d. K. D. Glinka

5. The process which operates in hydromorphic soils is known as

   a. Gleying b. Drying

   c. Vetting d. Water logging

6. The arable land in India is

   a. 170 million hectare b. 205 million hectare

   c. 159 million hectare d. 185 million hectare

7. The saline and alkali soils fall in which soil order

   a. Orthids and Argids b. Albolls and Xeroils

   c. Aqualts and Humults d. Andepts and Ochrepts

8. The Indian Institute of Soil Science is situated at
   a. New Delhi
   b. Bhopal
   c. Nagpur
   d. Indore
9. Which types of soils are covering largest area of India
   a. Laterite soil
   b. Deseret soil
   c. Alluvial soils
   d. Black cotton soils
10. Choose correct statement
    a. Minerals are organic in nature
    b. Minerals are both organic & inorganic in nature
    c. Minerals are inorganic in nature
    d. All of the above
11. Which is not matched with laterization?
    a. High temperature and high rainfall
    b. Removal of Fe and Al
    c. Tropical and semitropical region
    d. Formation of 1:1 type of clay mineral
12. In the designation of soil horizon describe capital latter O, L, A, E, C, R and W are represent of
    a. Solum
    b. Regolith
    c. Pedon
    d. Master horizon
13. Which one of the soil order is observed in all over India, except in hot and arid regions
    a. Aridisols
    b. Entisols
    c. Inceptisols
    d. Vertisols
14. The solum is comprised of
    a. A, B and E horizons
    b. A and B horizons
    c. A, B and C horizons
    d. A + regolith

15. The term 'Laterite' used by Buchanan in 1807 for typically formed soil in the climate condition of
    a. Alternate wet and dry seasons
    b. Continually waterlogging
    c. Continually dry
    d. Temperate climate

16. The dark black colour properties of vertisols due to
    a. Clay humus complex
    b. Titaniferous magnetite
    c. Iron and aluminium oxides
    d. Both (a) and (b)

17. Which is not under diagnostic subsurface horizons
    a. Argillic
    b. Natric
    c. Cambic
    d. Plaggen

18. Pyrite is
    a. Iron bearing mineral
    b. Iron and sulfur bearing mineral
    c. Sulfur bearing mineral
    d. None of the above

19. Mohr's scale is used to measure the
    a. Colour of mineral
    b. Light transmission
    c. Hardness of mineral
    d. Fracture

20. What is the lowest unit of soil classification ?
    a. Family
    b. Type
    c. Phase
    d. Series

21. The fine crystalline rock is
    a. Basalt
    b. Granite
    c. Pumice stone
    d. Gabbro

22. The comprehensive system of soil classification is based on
    a. Measurable soil properties
    b. Zonality concept
    c. Soil forming factors
    d. Organic matter content

23. Irregular rock masses formed in the preexisting cracks & crevices are
    a. Dykes
    b. Veins
    c. Sills
    d. Schiest
24. As per USDA Soil Taxonomy, soils have been grouped in to....... soil orders at the highest level of category
    a. 10
    b. 11
    c. 12
    d. 13
25. Ultra basic rocks contain
    a. Less than 55% silica
    b. Less than 65% silica
    c. Less than 45% silica
    d. Less than 75% silica
26. Stratification is a common feature in
    a. Igneous rocks
    b. Metamorphic rocks
    c. Sedimentary Rocks
    d. All above
27. The intermediate product formed in soil formation is
    a. Transformed material
    b. Residual material
    c. Parent material
    d. All above
28. ....rocks are the result of drastic changes brought about by heat and pressure
    a. Dolomite
    b. Marl
    c. Mica
    d. Quartz
29. Weathering is essentially a process of
    a. Disintegration
    b. Disintegration and decomposition
    c. Decomposition
    d. None of above
30. The great soil group of genetic system of soil classification are in azonal soils:
    a. Lithosols
    b. Alluvial soil
    c. Regosols
    d. All of these

31. Peeling of rook due to effect of temperature is

a. Exfoliation
b. Pneumatification
c. Foliation
d. Sterilization

32. Rocks formed by solidification of molten magma at the surface of the earth are called

a. Sedimentary
b. Metamorphic
c. Igneous
d. None

33. The quantitative relationship between soil forming factors and soil property is given by

a. Jenny
b. Liebig
c. Siribisteiv
d. Glinka

34. Which of the following soil order is not found in India ?

a. Mollisol
b. Vertisol
c. Gelisol
d. Inceptisol

35. Soil solum consist of horizon

a. A, B and C
b. E, A and B
c. O, A, B and C
d. B and C

36. Surface diagnostic horizon is called as

a. Endopedon
b. Regolith
c. Epipedon
d. All are correct

37. Least resistant mineral to weathering is

a. Quartz
b. Olivine
c. Geothite
d. Anorthite

38. Subsurface diagnostic horizon is called as

a. Epipedon
b. Endopedon
c. Solum
d. Ochric

39. In tropical climate the rate of weathering is

a. Lower than temperate climate
b. Higher than temperate climate
c. More of less same
d. None of above

40. Vertical section of a soil is known as

a. Horizons
b. Solum
c. Profile
d. A soil

41. The basic igneous rook generally produce

a. Coarse textured soil
b. Medium textured soil
c. Fine textured soil
d. All above soils

42. Soils developed from volcanic material and amosphous clay belong to order

a. Andisols
b. Gelisols
c. Spodosols
d. Vertisols

43. Pyroxene mineral are the example of

a. Tectosilicates
b. Phyllosilicates
c. Single-chain ionosilicates
d. Double-chain ionosilicates

44. Unconsolidated material above the bed rock is called

a. Solum
b. Regolith
c. A horizon
d. R horizon

45. The highest unit of soil classification taxonomy is

a. Order
b. Great group
c. Family
d. Series

46. Which one of the following is a Vertisol

a. Black cotton soil
b. Red sandy loam soil
c. Sandy loam sodic soil
d. Submontane (Tarai) soil

47. Shale rock develops from soil rich in

a. Sand
b. Loam
c. Silt
d. Clay

48. The B horizon is generally absent in

a. Black soils
b. Red soils
c. Desert soils
d. Brown hill soils

49. An illuvial horizon to clay accumulation is denoted by

a. A_p b. B_w

c. B_t d. B_h

50. A GIS is "a set of tools for collecting, storing, retrieving at will, transforming and displaying spatial data from the real world for a particular set of purposes" is a well used definition of a GIS provided by

a. Mike Goodchild (1997) b. Roger Chorley (1987)

c. David Rhind (1991) d. Peter Burrough (1986)

51. In land capability classification, lands suitable for cultivation are from class

a. I to III b. I to V

c. I to IV d. I to II

52. Which of the following is the correct sequence of element in the decreasing order of abundance in earth's crusts (46, 28, 8.3, 5.6%)

a. Si > O > Al > Fe b. Al > O > Fe > Si

c. O > Si > Al> Fe d. Fe > Si > Al > O

53. Soils formed from which of the following rocks are more fertile

a. Acid rocks b. Basic rocks

c. Neutral rocks d. Acid and neutral rocks

54. Unconsolidated material overlying rocks is known as

a. Solum b. Regolith

c. Bed rocks d. A horizon

55. Kaolinite is abundant in which of the following soil

a. Ultisols b. Inceptisols

c. Entisols d. Vertisols

56. Allophane material has been detected in which soil order?

a. Histosols b. Aridisols

c. Andisols d. Ultisols

57. Which of the following is the predominant mineral present in shale
    a. Calcite
    b. Dolomite
    c. Quartz
    d. Clay minerals
58. Soil colour inherited from its parent material is termed as
    a. Idiochromatic
    b. Pedochromatic
    c. Lithochromatic
    d. Stoichiometric
59. In an average, feldspar constitute how much percentage of igneous rocks by weight
    a. 60%
    b. 10%
    c. 5%
    d. 20%
60. The solubility of tourmaline is maximum in
    a. Acid soil
    b. Sodic soil
    c. Saline soil
    d. Organic soil
61. Total number of master horizons present in soil are
    a. Four
    b. Five
    c. Six
    d. Eight
62. The characteristics of Andisols are
    a. Soils developed from volcanic ash
    b. Weekly developed mineral soils with no profile development
    c. Soils with the initial stage of profile development
    d. Highly weathered, presence of oxic horizon
63. How does living plants appear on false-colour infrared images
    a. White
    b. Blue
    c. Black
    d. Red
64. Presence of argillic horizon with low base saturation found in which soil order?
    a. Inceptisols
    b. Spodosols
    c. Vertisols
    d. Ultisols

65. Soil order characterized by excess of weathering with dominant iron and aluinium oxide clays, common in tropics is

a. Entisols b. Mollisols

c. Alfisols d. Oxisols

66. A silicate-clay-enriched subsurface formed by illuviation of clay is known as

a. Agric b. Natric

c. Argillic d. Oxic

67. Lowest category of soil classification includes

a. Order b. Great group

c. Series d. Family

68. Who consider the various properties of soils in relation to plant production ?

a. Agronomists b. Edaphologists

c. Pedologist d. Soil chemists

69. In the category of family of soil taxonomy, major emphasis is given on

a. Diagnostic horizon b. Morphology

c. Heretic homogeneity d. Important properties for plant growth

70. Which soil order derived from volcanic ejecta and dominated by allophane or Al - humic complexes

a. Histosols b. Andisols

c. Ultisols d. Alfisols

71. According to the soil taxonomy desert soils belong to which of the following soils ?

a. Vertisols b. Spodosols

c. Inceptisols d. Aridisols

72. Orterde is a humus rich B-horizon found in

a. Chernozems b. Podzols

c. Gleisoils d. Rendzinas

73. Transformation of minerals during weathering follows the order?
   a. Mica⇌Illite⇌Vermiculite⇌Montmorillonite
   b. Mica⇌Montmorillonite⇌Vermiculite⇌Illite
   c. Mica⇌Vermiculite⇌Illite⇌Montmorillonite
   d. Mica⇌Montmorillonite⇌Illite⇌Vermiculite
74. Which one of the following minerals would weather most rapidly
   a. Muscovite
   b. Albite
   c. Quartz
   d. Microline
75. Argillic subsurface diagnostic horizon is a diagnostic feature of :
   a. Black soils
   b. Laterite and lateritic soils
   c. Red soils
   d. Alluvial soils
76. Dark coloured soils are the characteristics of
   a. Spodosols
   b. Entisols
   c. Ultisols
   d. Mollisols
77. Which suborder is not occurred under Entisols
   a. Aquents
   b. Orthents
   c. Fluvents
   d. Arents
78. Black soil commonly represent the order of
   a. Spodosols
   b. Entisols
   c. Ultisols
   d. Vertisols
79. How many diagnostic surface horizons consist in soil taxonomy
   a. 12
   b. 10
   c. 9
   d. 5
80. Mg-pyroxenes are more stable than which of the following minerals
   a. Biotite
   b. Amphiboles
   c. Ca-pyroxenes
   d. Olivines

81. Intermixing of soil horizons in certain soils has been termed

a. Horizonation
b. Gleization
c. Pedoturbation
d. Argillation

82. Which of the following are the passive factors of soil formation

a. Parent material and relief
b. Relief and climate
c. Relief and organisms
d. Relief and time

83. Rocks on soil debris accumulated at the part of slope due to gravity is known as

a. Lacustrine
b. Alluvial
c. Fluvial
d. Colluvial

84. What type of rock is likely to form when magma cools

a. Metamorphic
b. Igneous
c. Sedimentary
d. None of the above

85. Which of these would not cause weathering

a. Rain
b. Cloud
c. Sun
d. Frost

86. Identify the volcanic rock among the following:

a. Granite
b. Gabbro
c. Syenite
d. Rhyolite

87. A surface horizon with base saturation <50% and high C:N ratio, which is not naturally dry for more than three months per year is known as

a. Umbric
b. Mollic
c. Ochric
d. Anthropic

88. A sequence of soils of about the same age, derived from similar parent material, and occurring under similar climatic conditions, but having different characteristic due to variation in relief and drainage is called

a. Ca^{+} clays
b. Catena
c. Truncated profile
d. Caliche

89. Soils are acidic due to
    a. Low rainfall
    b. Acidic parent material
    c. Undulating topography
    d. None of these
90. Which horizon has maximum leaching and washing out
    a. A horizon
    b. B horizon
    c. C horizon
    d. Regolith
91. Soft and unconsolidated calcium carbonate, usually mixed with varying amounts of clay or other impurities is described as
    a. Mar
    b. Marl
    c. Mor
    d. Mull
92. In which type of survey, cadastral maps and aerial photographs are generally used as base material
    a. Detailed soil survey
    b. Reconnaissance soils survey
    c. Detailed-reconnaissance soils survey
    d. Semi-detailed soil survey
93. Which soil order stands for organic soils
    a. Histosols
    b. Inceptiosols
    c. Mollisols
    d. Oxysols
94. Soil order of highly weathered soils is
    a. Histosols
    b. Inceptiosols
    c. Mollisols
    d. Oxysols
95. The purity of the colour is soil represented by
    a. Hue
    b. Value
    c. Chroma
    d. None of these
96. Predominant clay mineral of spodosols is
    a. Kaolinite
    b. Montmorillonite
    c. Illite
    d. Sesquioxides

97. The weight of one hectare furrow slice (0-15 cm depth) is

a. $1.2x\ 10^6$ kg
b. $4.2x10^6$ kg
c. $2.2x10^6$ kg
d. $3.3x10^6$ kg

98. Most easily weatherable mineral in soil is

a. Olivine
b. Hornblende
c. Biotite
d. Quartz

99. Soils formed from which of the following rocks are more fertile?

a. Acid rocks
b. Basic rocks
c. Neutral rocks
d. Acid and neutral rocks

100. Dickite belongs to which group of minerals?

a. Kaolinite
b. Mica
c. Smectite
d. Amorphous

101. A silicate clay enriched subsurface horizon formed by illuviation is known as

a. Agric
b. Natric
c. Argillic
d. Oxic

102. Which science assumes the status of petrology, pedography and pedogenesis ?

a. Soil Science
b. Edaphology
c. Pedology
d. Petrology

103. The smallest volume that can be called 'a soil' is

a. Pedon
b. Ped
c. Radon
d. Pans

104. Soils have no diagnostic pedogenic horizons are included in soil order

a. Histosols
b. Mollisols
c. Ultisols
d. Entisols

105. According to the soil taxonomy, desert soils of India belong to which of the following order

a. Vertisols
b. Spodosols
c. Inceptisols
d. Entisols

106. Soils form on volcanic ash belongs to which of the following order?

a. Gelisols
b. Andisols
c. Oxisols
d. Inceptisols

107. What is the main basis of distinction between mollic and umbric epipedon?

a. Organic matter
b. Base saturation
c. pH
d. CEC

108. Which type of soil survey use the lowest categories of taxonomic and mapping units, i.e. types and phases

a. Detailed survey
b. Reconnaissance survey
c. Detailed reconnaissance survey
d. None

109. Fluor apatite is a rich source of

a. Nitrogen
b. Phosphorus
c. Potassium
d. Molybdenum

110. The peeling off the plates from parent rock due to differential expansion and contraction of minerals as a result of changes in temperature, is called as

a. Meander
b. Exfoliation
c. Illuviation
d. Moraine

111. Which of the following combination of minerals is most abundant in the silt and sand fractions of soils?

a. Quartz and iron-oxide minerals
b. Quartz and calcite
c. Feldspars and mica
d. Quartz and feldspars

112. Most dominant mineral in earth crust is

a. Feldspar
b. Amphiboles
c. Quartz
d. Mica

113. Which of the following has the highest weather ability

a. Biotite
b. Apatite
c. Garnet
d. Zircon

114. Red soils are predominant in

a. Tamil Nadu and Karnataka
b. Assam and Tripura
c. Gujarat and Maharashtra
d. Kerala and West Bengal

115. Regur is the local name for

a. Black soils
b. Lateritic soils
c. Hill soils
d. Saline soils

116. A subsurface horizon that is cemented mostly by silica is known as

a. Duripan
b. Fragipan
c. Placic
d. Glossic

117. The term catena (a Latin word means a chain) was introduced by

a. Milne (1935)
b. Schofield (1935)
c. Mulder (1909)
d. Richard (1950)

118. In soil taxonomy, diagnostic horizons are used to differentiate

a. Order
b. Sub-order
c. Great group
d. Series

119. Which of the following is the correct weathering sequence of mica?

a. Biotite → chlorite → vermiculite → smectite
b. Chlorite → biotite→ vermiculite → smectite
c. Vermiculite → chlorite → biotite → smectite
d. Biotite → chlorite → smectite→ vermiculite

120. Select the correct sequence of weathering stages in soil formation
   a. Juvenile → initial → virile → senile → final
   b. Senile → virile → juvenile → initial → final
   c. Initial → juvenile → virile → senile → final
   d. Initial → virile → juvenile → senile → final

121. Identify the volcanic rock among the following
   a. Rhyolite b. Granite
   c. Sylvinite d. Gabbro

122. Diagnostic horizon of a soil to be classified as Inceptisol is
   a. Cambic b. Oxic
   c. Spodic d. Argillic

123. The rate of weathering is generally high in
   a. Arid tropical climate b. Arid temperate climate
   c. Humid tropical climate d. Humid temperate climate

124. Solodization refers to the removal of..... from the exchange sites
   a. Ca^{++} b. Mg^{++}
   c. Na^{+} d. NaCl

125. Hydromorphic soils are evolved by which process
   a. Calcification b. Gleization
   c. Salinization d. Solonization

126. Which is not the passive factors of soil formations
   a. Parent material b. Climate
   c. Time d. Relief

127. The dominant mineral of sand, fine sand and course silt soils
   a. Pyroxine b. Feldspar
   c. Mica d. Quartz

128. In master horizon of a soil profile, which one is the horizon of accumulation

a. A horizon
b. E horizon
c. B horizon
d. C horizon

129. Soil mapping units in detailed soil survey of 1:5000 scales are

a. Sub group Association
b. Series Association
c. Phases of Soil Series
d. Family Association

130. When the mean annual soil temperature ranges between 8 °C to 15 °C, then the soil thermal regime are classified as

a. Frigid
b. Thermic
c. Mesic
d. Cryic

131. Evaluation of land for land use planning is a consequent step following

a. Soil conservation
b. Soil testing
c. Soil survey and mapping
d. Soil quality

132. Which process is responsible for accumulation of materials in B horizons?

a. Eluviation
b. Illuviation
c. Carbonation
d. Oxidation

133. Science dealing with the origin or formation of soil from the unconsolidated parent material with special reference to factors and process of soil formation is known as

a. Soil science
b. Pedology
c. Soil genesis
d. Edaphology

134. ....is the smallest unit of individual soil body which can be thick enough to indicate the whole profile down to the boundary with 'Not-soil'.

a. Endopedon
b. Epidedon
c. Pedon
d. Polypedon

135. Which of the following is active factor of soil formation ?

a. Parent material
b. Climate
c. Time
d. Relief

136. Which of the following elements has highest concentration in earth's crust ?

a. Iron | b. Silicon
c. Magnesium | d. Calcium

137. Among primary minerals, most resistant to weathering is

a. Quartz | b. Olivine
c. Albite | d. Gypsum

138. A boundary between soil and underlying coherent material, with a hardness of >3 on Mohr's scale is called..... contact

a. Paralithic | b. Lithic
c. Petrogypsic | d. Petroferric

139. Horizon enriched with Fe and Al oxides with dominance of 1:1 type clay minerals and from where silica has leached qualified for..... horizon

a. Kandic | b. Oxic
c. Sombric | d. Spodic

140. High sodium clay enriched horizon with columnar or prismatic structure qualifies for ... horizon

a. Agric | b. Kandic
c. Argillic | d. Natric

141. The smallest volume of soil that can be recognized as a soil individual is called as :

a. Pedon | b. Polypedon
c. Soil Profile | d. Soil Horizon

142. Which of following is a metamorphic rock ?

a. Sandstone | b. Dolomite
c. Granite | d. Gneiss

143. Sandstone and limestone are examples of

a. Sedimentary rocks | b. Igneous rocks
c. Metamorphic rocks | d. Mixed rocks

144. Which of the followings is a secondary mineral ?

a. Muscovite
b. Anorthite
c. Dolomite
d. Biotite

145. ....belongs to secondary minerals

a. Albite
b. Gibbsite
c. Biotite
d. Orthclase

146. How many number of classes consist in land capability classification ?

a. 8
b. 4
c. 6
d. 10

147. Which of the following is an acid igneous rock?

a. Basalt
b. Graphite
c. Granite
d. Gypsum

148.Soil horizons A, B and C (A+ B + C) are collectively called as :

a. Pedon
b. Soil profile
c. Soil regolith
d. Soil solum

149. Pedoturbation is the process of refer to

a. Mixing of the soil
b. Mixing of sand in soil
c. Mixing of fertilizer in the soil
d. Mixing of pesticides

150. Study of morphometric features and other characteristics of soil profile is

a. Pedogenesis
b. Pedography
c. Pedology
d. Pedoturbation

151. A thin, slowly permeable, dark reddish brown to black coloured iron or manganese pan that lies within 50 cm of the surface in a soil profile is termed as horizon

a. Albic
b. Kandic
c. Oxic
d. Placic

152. Who coined the term "pedology"

a. Friedrich Albert Fallou
b. Curtis F. Marbut
c. Justus von Liebig
d. Vasily V. Dokuchaev

153. The features on aerial photographs can best be interpreted by the use of

a. Kladioscope
b. Microscope
c. Stereoscope
d. GIS

154. Rocks formed by cooling of molten materials (magma) is called as:

a. Metamorphic rocks
b. Igneous rocks
c. Sedimentary rocks
d. Secondary rocks

155. Development of soil by physical weathering known as

a. Skeletal soils
b. Kankal soils
c. Kari soils
d. Both (a) and (b)

156. Materials transported and deposited by water are called as:

a. Alluvium
b. Colluvium
c. Loess
d. Aeolian

157. Soil order that is dominantly organic in nature is :

a. Mollisols
b. Spodosols
c. Histosols
d. Andisols

158.Choose the incorrect statement

a. The black soil of India formed by basaltic rocks
b. Podzolisation process is the opposite of calcification
c. The natural process of soil mixing termed as piping
d. Red & lateritic soil of India has been classified as alfisols

159. Inorganic parent material transported by gravity is

a. Glacial
b. Eolian
c. Marine
d. Colluvial

160. .... be longs to primary minerals

a. Calcite
b. Limonite
c. Kaolinite
d. Muscovite

161. Soil and Land Use Survey of India was established in 1958. Where is it not regional center situated?

a. Noida
b. Raipur
c. Nagpur
d. Bengaluru

162. Which one of the following soil orders represents old soils of humid region having low base status and clay enriched Bt horizon

a. Ultisol
b. Entisol
c. Alfisol
d. Mollisol

163. Zircon is............weatherable mineral

a. Easily
b. Moderately
c. Slowly
d. Strongly

164. Hardness of Quartz on Moh's scale is

a. 2
b. 7
c. 8
d. 6

165. Which of the following is the component of loamy soils?

a. Red soil
b. Clay soil
c. Alluvial soil
d. Sandy soil

166. Slickensides are the identifiable feature of soils classified as

a. Alfisols
b. Vertisols
c. Ultisols
d. Mollisols

167. COLE value is used as a parameter for characterization of soil classified under

a. Entisols
b. Oxisols
c. Histosols
d. Vertisols

168. Which one of the soil order having highest area in the world?

a. Vertisols
b. Andisols
c. Aridisols
d. Gelisols

169. Which one of the following is not a Mn mineral

a. Pyrolusite
b. Hausmanite
c. Rhodonite
d. Tenorite

170. The pedogenic process through which iron and manganese are reduced under anaerobic conditions is called

a. Humification
b. Laterization
c. Elluviation
d. Gleization

171. The inter-lattice swelling varies in clay minerals in the order

a. Vermiculite>Montmorillonite>Beidellite>Illite>Kaolinite>Halloysite
b. Vermiculite>Montmorillonite>Illite>Beidellite>Kaolinite>Halloysite
c. Vermiculite>Illite>Beidellite>Montmorillonite>Kaolinite>Halloysite
d. Vermiculite>Beidellite>Montmorillonite>Illite>Kaolinite>Halloysite

172. Choose the incorrect statement

a. Soil monoliths are permanent record of soil profile
b. Most of indo-gangetic plains are alluvial terraces.
c. Humification is the process of decomposition organic matter
d. Hydrolysis is the important processes of physical weathering

173. Which of the rock types listed below falls in the category of metamorphic rocks

a. Quartzite
b. Shale
c. Basalt
d. Syenite

174. An area on land represented by 1 cm^2 on 1:1000,000 scale map is

a. 50 ha
b. 500 ha
c. 100 ha
d. 1000 ha

175. In terms of the order from youngest to the oldest soil, which one of the following is the correct chronological sequence ?

a. Oxisol - inceptisol alfisol - entisol
b. Entisol - inceptisol - alfisol - oxisol
c. Entisol - alfisol - inceptisol - oxisol
d. Inceptisol - entisol - oxisol - alfisol

176. Which one of the following sets shows the correct soil order in sequence of increasing age

a. Entisol, mollisol, alfisol, ulfisol

b. Entisol, mollisol, ultisol, alfisol

c. Ultisol, alfisol, mollisol, entisol

d. Alfisol, mollisol, entisol, ultisol

177. The correct chronological sequence of soil orders in terms of their maturity is

a. Entisols-Inceptisols-Alfisols-Ultisols

b. Entisols-Inceptisols-Ultisols-Alfisols

c. Ultisols-Alfisols-Inceptisols-Entisols

d. Alfisols-Inceptisols-Entisols-Ultisols

178. .... is the process that removes silica, instead of sesquioxides, from the upper soil layers there by leaving sesquioxides to concentrate in the solum.

a. Calcification
b. Laterization
c. Podzolization
d. Salinization

179. Which one of the soil is known as chalka soil

a. Black cotton soil
b. Dark silty loam
c. Red sandy loam
d. White saline soil

180. Chernozem soils are found in

a. India
b. USA
c. Russia
d. Brazil

181. As per Goldich's stability series, the Olivines are

a. Least weatherable
b. Most weatherable
c. Moderately weatherable
d. Less weatherable than biotite

182. Sand and silt commonly washed away by flowing water in a process called as

a. Alluvium
b. Piping
c. Colluvium
d. Aeolian

183. Lithosols, Alluvial soils and Regosols of the genetic system of soil classification are grouped under the order

a. Zonal b. Azonal

c. Inter zonal d. Transitional

184. Silicious materials having poor reserves of weatherable minerals favour in operation of

a. Calcification b. Podzolization

c. Laterization d. Hydration

185. Which of the following remote sensing satellites was not launched by India

a. IRS 1D b. LANDSAT

c. CARTOSAT d. RESOURCESAT

186. Which one of the soil order is mostly classified as red soil

a. Ultiisols b. Inceptisols

c. Alfisols d. Andisols

187. The "International year of soil" by UNO celebrated in the year

a. 2005 b. 2015

c. 2012 d. 2018

188. The science of study of earth's crust is known as:

a. Mineralogy b. Crystallography

c. Petrology d. Geology

189. Albite is the pure minerals of

a. Calcium b. Magnesium

c. Potassium d. Sodium

190. World Soil Day is celebrated every year on

a. 16th September b. 16th October

c. 5th November d. 5th December

191. Which one of the following is a manganese containing minerals

a. Magnetite
b. Tourmaline
c. Apatite
d. Pyrolusite

192. The most abundant non-metal elements in the earth's crust are

a. Silica and aluminium
b. Silica and oxygen
c. Silica and calcium
d. Silica and magnesium

193. The softest mineral in the earth crust is

a. Gold
b. Quartz
c. Diamond
d. Talc

194. Peeling of rock due to effect of temperature is

a. Exfoliation
b. Pneumatification
c. Foliation
d. Sterilization

195. Which one of the following is the primary mineral that occurs predominantly in igneous rocks

a. Micas
b. Feldspars
c. Quarty
d. Pyroxenes

196. Average annual loss of soils due to erosion is

a. 2.5 billion tonnes
b. 5.0 billion tonnes
c. 7.5 billion tonnes
d. 9.0 billion tonnes

197. To which one of the soil orders do black soils of India belong

a. Alfisol
b. Oxisol
c. Inceptisol
d. Vertisol

198. The most advanced stage of weathering is represented by

a. Entisols
b. Oxisols
c. Gelisols
d. Histosols

199. Average content of silica in soil is

a. 24%
b. 27 %
c. 31 %
d. 35 %

200. A soil profile having two sola, one above the other is known as

a. Bisequem profile
b. Truncated profile
c. Soil monolith
d. Soil fabric

201. Among the following sources of salts in soils, which one is the primary source

a. Ground water
b. Organic manures
c. Canal irrigation
d. Rock weathering

202. The marble is an example of

a. Metamorphic rock
b. Sedimentary rock
c. Igneous rock
d. Soil minerals

203. Which soil order have highest area in India

a. Entisols
b. Inceptisols
c. Vertisols
d. Alfisols

204. In India, which soil order have negligible area

a. Oxisols
b. Ustisols
c. Mollisols
d. Histosols

205. Gilgai micro relief is found in which order

a. Vertisol
b. Entisol
c. Andisol
d. Histosol

206.Eluviations refers to

a. Removal of clay from above horizons to lower horizons
b. Removal of clay and sesquioxide from above horizons to lower horizons
c. Removal of soluble salts from above horizons to lower horizons
d. Accumulation of clay and sesquioxide on the surface layer

207. Soils of desert lands are known as

a. Alfisols
b. Spodosols
c. Mollisols
d. Aridisols

208. Which rock is volcanic (extrusive) in native

a. Gabbro
b. Diorite
c. Basalt
d. Monzonite

209. Rocks containing more than 65 % SiO_2 are

a. Neutral
b. Alkaline
c. Acidic
d. Calcareous

210. Which of the following is the amorphous clay mineral

a. Muscovite
b. Kaolinite
c. Vermiculite
d. Allophane

211. Which one is not a potash-bearing mineral

a. Feldspar
b. Olivine
c. Muscovite
d. Biotite

212. Division of the factors of soil formation in to active and passive groups was proposed by

a. H. Jenny (1941)
b. C.F. Marbut
c. J.S. Joffe (1949)
d. V.V. Dokuchaiev

213. Which one is not a pedogenic process

a. Humification
b. Calcification
c. Laterization
d. Ferruginization

214. A subsoil layer of high bulk density which is brittle when moist and very hard when dry, is known as

a. Fragipan
b. Duripan
c. Durinodes
d. Plinthite

215. Ultisols are comparable with Alfisols except for having

a. High base saturation of the exchange complex
b. Low base saturation of the exchange complex
c. Uniform soil profile
d. High water table

216. India is divided into how many Agro-climatic zones ?

a. 5
b. 10
c. 15
d. 20

217. India is divided into how many Agro-ecological regions ?

a. 5
b. 10
c. 15
d. 20

218. Total number of agro-ecological sub-region (AESR) in India delineated by NBSS&LUP is

a. 21
b. 42
c. 60
d. 76

219. In Soil Taxonomy, great groups are defined largely by

a. Soil forming processes
b. Presence or absence of diagnostic horizons
c. Thermal regime
d. Moisture regime

220. Which one is not a process of chemical weathering

a. Hydration
b. Hydrolysis
c. Precipitation
d. Carbonation

221. In the Jenny's factors of soil formation, factor 'b' refers to

a. Biological factor
b. Biochemical factor
c. Base-exchange
d. Biosphere

222. Sediments deposited in fresh water lakes are called

a. Delta
b. Aeolian
c. Lacustrine
d. Outwash

223. How many number of diagnostic surface horizons (epipedons) included in Soil Taxonomy

a. 6
b. 9
c. 12
d. 19

224. Argillic horizon is rich in

a. Fe and Al-oxides
b. Sodium
c. $CaCO_3$
d. Silicate clay

225. The Journal Communications in Soil Science and Plant Analysis is published by

a. Elsevier
b. Kluwer
c. CSIRO
d. Francis and Taylor

226. The diagnostic surface horizons are called

a. Genetic horizons
b. Pedons
c. Epipedons
d. Endopendons

227. Podzolization is a process accumulation in soil

a. Silica
b. Iron and aluminium oxides
c. $CaCO_3$
d. $CaSO_4$

228. Laterization is a process of accumulation in soil

a. Iron and aluminium oxides
b. Silica
c. $CaCO_3$
d. $CaSO_4$

229. Which mineral substance placed at number 10 in Mho's scale

a. Talc
b. Gypsum
c. Carborundum
d. Diamond

230. Aridisols include most of the soil of arid region, those formerly designated as

1. Red loams
2. Grey brown
3. Solonchak
4. Calcareous sierozem

Select the correct answer using the codes:

a. (1), (2) and (4)
b. (2) and (4)
c. (2), (3) and (4)
d. (3) and (4)

231. Francolite contains

1. Fluoroapatite
2. Chloroapatite
3. Hydroxy apatite
4. Carbonate apatite

Select the correct answer using the codes:

a. (1) and (2) b. (2),(3) and (4)

c. (1) and (4) d. (3) and (4)

232. The Q/I relationship of soil phosphorus provides information on

1. Immediately available phosphorus
2. Ionic balance
3. Phosphate buffering capacity of soil
4. Interference of other nutrients

Select the correct answer using the codes:

a. (1), (2) and (3) b. (2) and (3)

c. (2) and (4) d. (1) and (3)

233. Which of the following are passive soil forming factors

1. Topography 2. Time

3. Biosphere 4. Parent material

Select the correct answer using the codes:

a. (1) and (2) b. (1) and (4)

c. (1), (2) and (4) d. (1), (3) and (4)

234. Match List I (Soil order) with List II

a. Aridisols	1. Turn over
b. Entisols	2. Organic
c. Alfisols	3. Arid
d. Histosols	4. Aluminium and Iron
e. Vertisols	5. Recent

235. Match List I with List II

a. Primary mineral	1. Mica
b. Secondary mineral	2. Metamorphic rock
c. Silica	3. Greyish tinge to soil
d. Limestone	4. Gypsum
e. Marble	5. Sedimentary rock

236. Match List I with List II

a. Igneous rock	1. Shale
b. Mineral	2. Granite
c. Clay	3. Illite
d. Metamorphic rock	4. Gneiss
e. Sedimentary rock	5. Diamond

237. Match List I with List II

a. Permafrost	1. Spodosols
b. Slickensides	2. Andisols
c. Low bases	3. Vertisols
d. Amorphous material	4. Gelisols
e. Albic horizon	5. Ultisols

238. Match List I with List II

a. Cemented by silica	1. Albie
b. Light coloured A_2 horizon	2. Duripan
c. Accumulation of low activity clays	3. Kandic
d. Fe and Mn pan	4. Placic

239. Match List I with List II

a. Climatic classification	1. Guy D Smith
b. Stability index of minerals	2. Dokuchaev
c. Comprehensive system of soil classification	3. Vilenski
d. Zonality concept	4. Jackson

ANSWERS (Pedology)									
1.c	2.c	3.d	4.c	5.a	6.c	7.a	8.b	9.c	10.c
11.b	12.d	13.c	14.a	15.a	16.d	17.d	18.b	19.c	20.d
21.a	22.a	23.b	24.c	25.c	26.c	27.c	28.c	29.c	30.d
31.a	32.c	33.a	34.c	35.b	36.c	37.b	38.b	39.b	40.c
41.c	42.a	43.c	44.b	45.a	46.a	47.a	48.a	49.c	50.a
51.c	52.c	53.a	54.b	55.a	56.c	57.c	58.c	59.a	60.a
61.b	62.a	63.d	64.d	65.d	66.c	67.c	68.b	69.d	70.b
71.d	72.b	73.a	74.b	75.a	76.d	77.d	78.d	79.c	80.d
81.c	82.a	83.d	84.b	85.b	86.d	87.a	88.b	89.b	90.a
91.b	92.a	93.a	94.d	95.b	96.d	97.c	98.a	99.b	100.a
101.c	102.c	103.a	104.d	105.d	106.b	107.b	108.a	109.b	110.b
111.d	112.a	113.a	114.a	115.a	116.a	117.a	118.c	119.a	120.c
121.a	122.a	123.c	124.c	125.b	126.b	127.b	128.c	129.c	130.c
131.c	132.b	133.c	134.c	135.b	136.b	137.a	138.b	139.b	140.d
141.a	142.d	143.c	144.c	145.b	146.a	147.c	148.c	149.a	150.b
151.d	152.a	153.d	154.b	155.d	156.a	157.c	158.c	159.d	160.d
161.b	162.a	163.c	164.b	165.c	166.b	167.b	168.c	169.d	170.d
171.a	172.d	173.a	174.d	175.b	176.a	177.a	178.b	179.c	180.c
181.d	182.b	183.b	184.b	185.b	186.c	187.b	188.c	189.d	190.d
191.d	192.b	193.d	194.a	195.b	196.b	197.d	198.b	199.b	200.a
201.d	202.a	203.b	204.d	205.a	206.b	207.d	208.c	209.c	210.d
211.b	212.c	213.b	214.a	215.b	216.c	217.d	218.c	219.b	220.c
221.d	222.c	223.b	224.d	225.d	226.c	227.a	228.a	229.d	230.b
231.c	232.d	233.c							
234. a.3 b.5 c.4 d.2 e.1					235. a.4 b.1 c.3 d.5 e.2				
236. a.2 b.5 c.3 d.4 e.1					237. a.4 b.3 c.5 d.2 e.1				
238. a.2, b.1 c.3 d.4					239. a.3 b.4 c.1 d.2				